Ornaments and surfaces

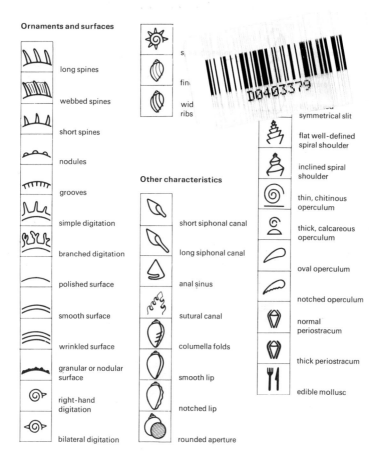

long spines

webbed spines

short spines

nodules

grooves

simple digitation

branched digitation

polished surface

smooth surface

wrinkled surface

granular or nodular surface

right-hand digitation

bilateral digitation

s

fin

wid
ribs

Other characteristics

short siphonal canal

long siphonal canal

anal sinus

sutural canal

columella folds

smooth lip

notched lip

rounded aperture

symmetrical slit

flat well-defined spiral shoulder

inclined spiral shoulder

thin, chitinous operculum

thick, calcareous operculum

oval operculum

notched operculum

normal periostracum

thick periostracum

edible mollusc

Relative market rarity in relation to price in US$

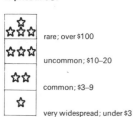

rare; over $100

uncommon; $10–20

common; $3–9

very widespread; under $3

Depth at which shells are to be found

Metres	Fathoms	
m	FATH	
+2	+1	up to 2m above low-water mark
0	0	at low-water mark
-2	-1,1	about 2m
-5	-2,5	about 5m
-10	-5,5	about 10m
-20	-11	about 20m
-30	-16	about 30m
-50	-27	about 50m
-70	-38	about 70m
-100	-55	about 100m
-150	-82	about 150m
↓	↓	about 200m

Shells
a collector's
colour guide

J. & R. Senders

English translation by Lucia Wildt

DAVID & CHARLES
Newton Abbot   London

All the illustrations in this book are from the authors' collection and have been taken by them.
Cover illustration: *Pleurotomaria teramachii* Teramachi's slit shell

British Library Cataloguing in Publication Data

Senders, J.
  Shells.
  1. Molluscs—Collection and preservation
  2. Shells
  I. Title   II. Senders, R.
  III. Guide des coquillages de collection.
  English
  594′.0471   QL.406.5
  ISBN 0–7153–8497–X

Filmset by Keyspools Ltd, Golborne, Lancs
Printed in Belgium

CONTENTS

INTRODUCTION

Like the other volumes in this series, this particular book is a practical guide aimed at the inexperienced as well as the specialist collector – indeed all who wish to get to know shells, from the simple souvenir to the spectacular specimen. It is packed with practical information in a form which will save much time in identifying shells and avoid inconvenience and confusion.

Since collecting shells should always remain a hobby, a spare-time occupation – sometimes even a passion – we have systematically avoided rare specimens commanding high prices. The only exceptions to this rule can be justified on the grounds that they are widely known: the *Conus gloriamaris*, renowned both for its name and its beauty, or the *Cypraea aurantium*, the golden cowry, whose unique colour made it the royal symbol of the chieftains of the Fiji islands; or on grounds of biological peculiarities: the Pleurotomariidae which had only been known as fossils until discovered, alive, in the waters of Japan (they were reserved for the Emperor's personal collection until quite recently).

The coverage of this book is limited to the shells of warm waters because those we find on the shores of the North Atlantic, the English Channel and the North Sea are usually dead ones. The majority are bivalves, hardly identifiable to the eyes of the inexperienced, and the anatomical criteria which characterise them are not easily accessible to the novice. Tropical sea shells, on the other hand, come in extremely varied shapes: their splendid patterns and constantly different colours make them much more easily identifiable. Finally, notwithstanding the practical nature of this guide, we wished to play a part in the reader's dreams: little imagination will be needed to follow the fisherman or the diver into that underwater world where everything is at once beautiful and strange, or to find oneself strolling along endless white beaches accompanied only by the murmur of the waves.

Helped by this book, with its 144 colour plates and pictograms, the reader will quickly be able to establish the family to which any given specimen belongs. This information should be put onto a label which should accompany each shell or group of shells. In order to classify a shell with certainty, particularly when it comes to its species, sub-species and form, it is very important to know its exact provenance: bags and boxes containing shells should therefore be labelled with the place, the date and the means by which they were collected (on the beach, under corals, by hand, by diving, etc) as well as the name of the person who collected them. Any other detail, however irrelevant at first sight, could turn out to be important – even decisive should the classification be contested or the discovery of a new species be published.

The task of the beginner wishing to obtain a well-organised collection will be made easier if he belongs to a collectors' club or a group of malacologists (the latter study not only the shells but also the living organisms – the molluscs – which produce them). With this in mind, a section on collectors' clubs and associations is included (page 177). For a small annual subscription, members could attend meetings and classification sessions, gain advice from specialists (many collectors have their favourite families), have access to specialist books and the opportunity to acquire further specimens. Organised excursions would teach one how to recognise the biotopes most likely to yield shells in quantity, and one would be able to compare one's shells with those of expert collectors and in museums. Furthermore, a club magazine would announce a programme of meetings as well as providing numerous illustrations and descriptions of various species, including the latest discoveries from all corners of the world.

Before launching into the book, we would like to thank M. E. Brutsaert, Editorial Director of Editions Duculot at Gembloux, for his encouragement and dynamism which have supported us in our work. We would also like to thank M. R. Duchamps, President of the Belgian Malacological Society, for the fruitful consultations we have had with him, and we should not forget our correspondents from the four corners of the world who provided us with material and ideas. Over the years we have made friends with many of them, and although we cannot list them here, we are very grateful to them all.

NOMENCLATURE OF SHELLS

Today over 100,000 species of living molluscs are known; obviously they had to be classified according to different criteria and had to be given a name. Many tried their hand at this, but nomenclature was radically transformed by the publication in 1758 and 1767 of Linnaeus' *Systema Naturae*. The International Commission of Zoological Nomenclature (ICZN) has ratified the Linnaean binomial system and has established the Law of Priority according to which each species is called by the first name given it. The Latin terminations of each systematical group are the following:

-acea for the super-family   eg Mytilacea
-idae for the family   eg Mytilidae
-inae for the sub-family   eg Mytilinae

These are followed by the name of the genus or generic name (eg *Mytilus*); the name of the sub-genus is usually in parenthesis between the genus and the species. Above each plate for the sake of clarity is given the genus followed by the species; below each plate is given the sub-genus followed by the species; the name of the author and, when available, the year of the description; the family.

A shell's form often depends on its environment – thicker and smoother in a stormy habitat; thinner and more sculpted in a calm, deep environment. An informed collector can therefore derive pleasure from gathering shells of the same species but from different origins and habitats. He will thus be able to collect interesting information on the relationship between habitat and form and to realise, for instance, that in certain areas (north of Sri Lanka and Kwajalein in the Marshall Islands) shells' colours are much more constant. Simple discoveries such as these will probably help, one day, in establishing or confirming a scientific explanation yet to be found.

HOW TO READ THE SYMBOLS

Where molluscs live

on rocks, between the tide marks

buried in the sand between the tide marks, easier to find at low tide

under the overhang of rocks or coral reefs

in the branches of live corals

buried in the sand at the foot of rocks or corals

on *gorgonias*,* well camouflaged

on sandy areas covered in seaweeds

on the sand of the seabed, generally at night

buried in the sand of the seabed, mainly during the day

on top of *madrepores*, mainly at night

on the upper side of sponges, mainly at night

inside amphora-shaped sponges, mainly during the day

among tree roots in *mangrove* swamps

at the bottom of holes bored in wood

at the bottom of holes bored in rocks

* The explanation of words in italics can be found in the Glossary

How to collect them

 by hand

 by hand under the sand at low tide

 by shifting the sand with flippers at a depth of 3–4m

 with fishing groundlines in the case of certain carnivores

 accidentally in fishing nets meant for fish

 with specially designed fishing nets, sunk to a depth of up to 200m

 accidentally in cages designed to catch deepwater crustaceans

 with dredges specially designed to collect shells

 in the stomach of fish (some rare specimens have only been found this way)

Take care!

 The sting of this mollusc can be fatal

Shape and overall composition of the shell

 a shell formed by 8 plaques

 2 symmetrical valves joined by a *hinge*

 2 asymmetrical valves

 2 heart-shaped valves

 1 spiral-like valve with 5–7 orifices on the margin

 1 cap-shaped valve with 1 orifice on the tip

 1 more or less polygonal valve, cap-shaped

 horn-shaped shell, slightly curved

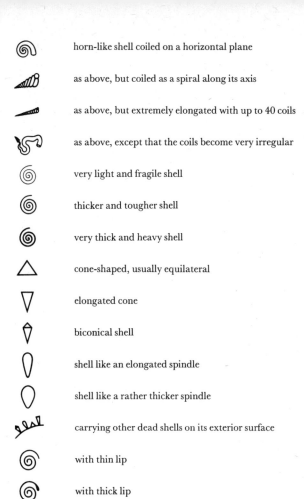

horn-like shell coiled on a horizontal plane

as above, but coiled as a spiral along its axis

as above, but extremely elongated with up to 40 coils

as above, except that the coils become very irregular

very light and fragile shell

thicker and tougher shell

very thick and heavy shell

cone-shaped, usually equilateral

elongated cone

biconical shell

shell like an elongated spindle

shell like a rather thicker spindle

carrying other dead shells on its exterior surface

with thin lip

with thick lip

with curled lip

Main patterns

transversal lines or bands

wavy transversal lines

longitudinal lines or bands

wavy longitudinal lines

spotted

large spots

ocelli

zigzag or marbled markings

square reticulations

rectangular reticulations

Ornaments and surface

long spines

webbed spines

short spines

nodules

grooves

simple *digitation*

branched *digitation*

polished surface

smooth surface

wrinkled surface

granular or nodular surface

right-hand *digitation*

bilateral *digitation*

digitations present on each spiral

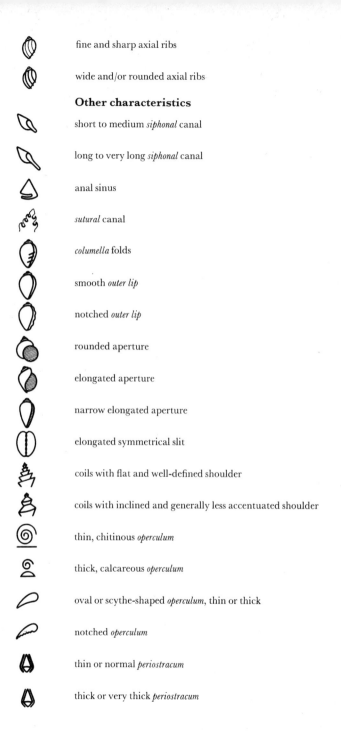

fine and sharp axial ribs

wide and/or rounded axial ribs

Other characteristics

short to medium *siphonal* canal

long to very long *siphonal* canal

anal sinus

sutural canal

columella folds

smooth *outer lip*

notched *outer lip*

rounded aperture

elongated aperture

narrow elongated aperture

elongated symmetrical slit

coils with flat and well-defined shoulder

coils with inclined and generally less accentuated shoulder

thin, chitinous *operculum*

thick, calcareous *operculum*

oval or scythe-shaped *operculum*, thin or thick

notched *operculum*

thin or normal *periostracum*

thick or very thick *periostracum*

🍴 edible mollusc

Relative market rarity in relation to price in US$

✩✩✩	rare	over $* 100
✩✩✩	uncommon	$ 10–20
✩✩	common	$ 3–9
✩	very widespread	below $ 3

* Price in US $ valid at the end of 1982; it should obviously be taken as an average and it always applies to a perfect specimen of medium size and normal colour and pattern. Prices are considerably lower for imperfect specimens – those with a broken tip, notched lip, or even a growth mark which, however natural, can spoil the appearance of the shell considerably.

Depth at which shells can usually be found

m	FATH	
+2	+1	in the area between the tide marks, up to 2m above low-water mark
0	0	on the tidal margin in very little water, or in the sand at low tide
-2	-1.1	2m
-5	-2.5	5m
-10	-5.5	10m
-20	-11	20m
-30	-16	30m
-50	-27	50m
-70	-38	70m
-100	-55	100m
-150	-82	150m
↓	↓	200m

Certain species have been found at depths lower than 200m.

The reader will find that some pictograms are incomplete: the biotope of some shells is still unknown and in such cases we preferred not to provide doubtful information. The reader can always add the colour himself should he be able to gain information from a reliable source. On the other hand, our pictograms have been conceived so as to allow their further use in other works on rare shells.

HISTORICAL INTRODUCTION

Since the dawn of time man has shown an interest in shells; food, domestic life, commercial activities, religious beliefs, fine art – there is hardly a field which has ignored them. We give here a résumé of the various ways in which shells have been used through the ages.

Molluscs have always formed an important part of man's diet. In our own countries, oysters, mussels and clams are very popular; turbinids, vermetids, lambis, scallops, and sea-ears are all used as ingredients in exotic dishes which are often very tasty. Having disposed of the animal, the shell was often wisely adapted to daily needs. Skill and ingenuity transformed the *Bursa* into lamps, the *Cymbium* into bailers to empty canoes, the *Cypraea tigris* into decoys for octopuses due to their fine glitter (Society Islands), the *Cypraecassis rufa* with its toothed lip into a cutting tool (Tuamotu), the *Cypraea mauritania* into a fruit grater (Marquesas), and the mother-of-pearl of some oysters into fish hooks (Polynesia). The giant clams with their wavy and ornamental shape could be used to display objects or, in the case of the very large ones, to hold water. Patiently filed down they were used in the construction of adzes, or *toki* (New Guinea). Specimens of *Charonia*, with the apex broken off, are still used by many peoples for religious ceremonies as well as to call to arms or to warn of danger. Perliferous oysters, quite common in Polynesia, were used as castanets during local festivities.

Quite early on man discovered the properties of certain shells such as the extraction of writing ink and dyeing pigments. Tyre purple, a monopoly of the Phoenicians who extracted it from *Murex brandaris* and *Trunculus*, was reserved for Roman aediles before being utilised by the Christian Church. During the Roman Empire the *byssus* of the nacre, woven by specialised artisans, provided a very tough, silky material which was highly prized and exported as far as Egypt. It was still used to make gloves in the Neapolitan area up to the First World War.

Since the nineteenth century perliferous oysters and *Trochus niloticus* have been exploited on an industrial scale: mother-of-pearl buttons were replaced by synthetic buttons only after 1950.

The *Cypraea moneta* owes its name to its widespread use, in the majority of African countries and in the Indian Ocean, as a currency – a centuries-old custom which lasted until the industrial era. Large quantities of this shell-coin came from the Indian Ocean and were taken to Liverpool (as a sideline of the slave market) to be used to pay for the goods which European merchants were importing from West Africa.

In California, sea-ears, cut into pieces, were also used as coins. The tribes of the north-west Pacific used to string the tusk shells to make garlands about two metres long, which were regarded as extremely valuable: they are called *haik-wa*. All along the eastern coast of North America and the commercial routes of the interior, the *wampum* was used as currency. The Indians would manufacture it with the fragments of a bivalve locally called *quahog* (*Mercenaria mercenaria Linnaeus*) or of a whelk (*Busycon carica* Gmelin). The small, shiny, well-polished cylinders were pierced and strung and mounted on belts; *wampums* were used in important commercial exchanges as well as to settle ransom demands.

Primitive peoples attributed magic powers to certain shells. Ancestral customs have changed over the centuries but some of those practices still survive. Ancient Egyptians used to place a cowry on the eyes of mummies to ensure they would retain their eyesight in the other world. In Zaire, the Basonga carry wooden fetishes adorned with cowries as a cure for abdominal troubles. The Tuaregs hang a fetish shell in their tents – a large cone, a cowry or a coral to keep evil spirits at bay.

Cowries, extremely prolific molluscs, are considered symbols of fertility: in parts of Asia women about to give birth hold a cowry in each hand to help the delivery. In Ghana from puberty onwards girls carry on their loins a puppet covered in cowries. In Tahiti, during funeral ceremonies, natives use oysters to cut their own flesh and smear themselves with blood to avenge the soul of the dead. Also in Tahiti, master tattooists used a small mallet and a finely sharpened nacre blade to break the skin.

Used as decorative elements, shells can be found on masks, hairstyles and ceremonial head-dresses, huts, totems and boats of many African tribes as well as in New Guinea and the Solomon Islands. The Greeks, the Romans and the Incas all adorned their terracottas with stylised shells. The Aztecs chose shells as pictographical adornments of their manuscripts.

Much later, shells supplied jewellers and goldsmiths with wonderful basic materials: the use of mother-of-pearl inlay flourished under the able fingers of seventeenth-century cabinet-makers. The precious cameos carved on *Cypraecassis rufa* and *Strombus gigas* are a speciality of Italian jewellery from the Renaissance to the present times. Convicts, from Devil's Island to Cayenne, have engraved and carved the shells of *Nautilus* to the point of achieving works of art as airy as lace.

The pearls produced by oysters have always fascinated men and women. A splendid piece of jewellery was created at the turn of this century by using several pearls, the largest of which – shaped like an aubergine – weighs 2400 grains, ie about 127g; it is the largest of all known pearls. This piece of jewellery is in the private collection of a rich Londoner.

Shells have also been used as religious symbols; examples include the scallop (*Pecten Jacobaeus*) carried by pilgrims to the shrine of St James of Compostela, the sacred chank (shell) of Hinduism (see plate 88), the heraldic motif of certain English families, such as the Churchills, whose ancestors took part in the Crusades.

The artists of the Italian Renaissance used shells in the representation of pagan as well as religious scenes: the *Birth of Venus* by Botticelli (1478) is the most famous example of this. The goddess, covered only by her long hair, emerges from the ocean supported by a giant scallop. In architecture, the shell has appeared in the gardens and grottoes of villas and castles from the sixteenth to the eighteenth century. Shells also figured largely in still-life paintings and the famous Savonnerie tapestries since the seventeenth century.

Finally, it is worth noting that more than fifty countries have adopted shells as a philatelic or numismatic motif, and that a well-known international oil company owes its name to its founder's interests as a shell collector and dealer.

This interest in shells gave rise, very early on, to the collecting craze: in AD40 the Emperor Caligula ordered his troops stationed along the coasts of France (ready to invade Britain) to collect as many shells as possible. Once back in Rome, he kept boasting about the precious hoard extracted from the ocean. During the seventeenth and eighteenth centuries 'curiosity' cabinets proliferated: they housed collections of strange and rare specimens from both the animal and the vegetal kingdoms. George Louis Leclerc de Buffon (1749–88) gave the King's 'Cabinet', which Louis XIII had started in 1635, a really scientific basis. In this period, European monarchs and leading figures (Catherine II of Russia, Louisa Ulrica, Queen of Sweden, Cosimo III, Grand Duke of Tuscany) vied with one another to acquire rare shells and spent large fortunes doing so.

Jean-Baptiste de Monet, Count of Lamarck (1744–1829), a zoology professor at the Natural History Museum in Paris, assembled a very rich collection of shells which allowed him to describe many new species. But the serious scientific explorations of the Pacific only started during the second half of the eighteenth century with Louis-Antoine de Bougainville and Captain Thomas Cook. The malacological discoveries occasioned by such expeditions are related in *The Universal Conchologist* (1784–7) by Martyn, a book illustrated by splendid colour pictures. Plate 39 shows the *Cypraea aurantium* discovered in Tongo in 1773. Only in 1791 did Gmelin give it its name according to Linnaeus' binomial system.

Hugh Cuming (1791–1865), an Englishman, three times sailed to the Pacific and his conchological observations were published by the Zoological Society in 1828. He also gathered a collection of about 19,000 species which has been housed in the British Museum (Natural History) since 1866.

One of the most important European collections was gathered by Philippe Dautzenberg (1848–1935), a Belgian businessman who earned a fortune in the carpet industry and devoted it to shells and conchology. His collection of 30,000 species and his prestigious library have been bequeathed to the Royal Institute for Natural Sciences in Brussels. One of the most important Japanese collections is the one belonging to the Emperor Hirohito, a marine biologist always profoundly interested in anything pertaining to the sea. We should also mention the rich collections of J. E. du Pont of Pennsylvania, numbering 300,000 specimens, and of the Smithsonian Institute, Washington, with more than 100,000.

The history of shell collecting is full of amusing and fascinating anecdotes to which no amateur can be indifferent. *Epitonium scalare*, for instance, was very rare and highly sought after in the eighteenth century. The specimen belonging to Catherine II was regarded as the largest. The shell was so highly prized that imitations were made with rice dough, which made them rather easy to detect: immersed in water, the false shells would collapse into a shapeless mass. Nowadays, anybody can easily buy specimens of this beautiful mollusc.

The *Conus milneedwardsi* is one of the most elegant as well as one of the rarest cones; only a dozen specimens were known in the nineteenth century, all from the coasts of East Africa at the latitude of the present Mozambique, and usually found dead on beaches where they had been thrown by the storms. One day, Captain F. W. Townsend found two of these molluscs attached to an underwater telegraphic cable which was being hauled from a depth of 100 metres for repairs. They were within reach of his hands when the larger one let go of the cable and dropped back into its own element.

The *Conus gloriamaris* was the most expensive shell in the world only a couple of centuries ago, and perhaps the rarest. The following anecdote is still recalled: a French collector owned one of the only two known specimens, the other being in the collection of the Dutchman Hwass. When the latter specimen came onto the market, the Frenchman bought it and smashed it with his heel: 'I now own the only specimen in the world!' he exclaimed. Times have changed: it is no longer necessary to be a magnate in order to own a *Conus gloriamaris*; each year, dozens of this fine shell are being fished in the Philippines alone, thanks to the ingenuity of the locals who have invented special lines which can descend to a depth of more than 100 metres. It is in this area that most of the recent discoveries (or re-discoveries) have been made.

Collectors have at their disposal a rich bibliography of over 4000 volumes and 90,000 abstracts devoted to shells. The first illustrated work was published in Rome in 1681: it was compiled by the Jesuit Filippo Buonanni and entitled *Recreatio mentis et oculi in observatione cochlearum* (*Entertainment of mind and eyes through the study of shells*). In 1685 Martin Lister published his *Historia conchyliorum* with plates well suited to the identification of many shells. However, the first relatively popular book on the subject was written in 1742 by the Frenchman Joseph Dellazier d'Argenville.

We thus come to the Swedish naturalist Carl von Linnaeus, whose binomial nomenclature is still in use; his book, *Systema naturae* (10th edition 1758), although devoting little space to the molluscs, is still to be regarded as fundamental. He was followed by the German painter Knorr who, impressed by the beauty of shells, published his *Delight of eyes and mind* in six volumes (1760–73) illustrated with wood engravings.

Commercial activities in this field grew apace with their rapidly increasing scientific curiosity. In 1830 Marcus Samuel founded a small transport company active between London and the East: he exported manufactured goods to the East and imported exotic eastern products. The firm grew rapidly and became the Shell Transport and Trading Company; its first ship was called *Murex*. The ships of the Royal Dutch Company, whose emblem is a scallop, are all named after a shell belonging to this family.

These days, countries like the Philippines, Taiwan, India and the Bahamas export yearly hundreds, and even thousands, of tons of shells, both cleaned and in their natural state. We have come a long way from the few hundred specimens sold yearly by a dealer specialising in collectable shells.

THE SHELLS

DEFINITION

A shell is the calcareous involucre of a mollusc. The zoological group of molluscs is a very important one and includes over 100,000 species; this makes it necessary to define molluscs on the basis of certain characteristics.

A mollusc's body consists of five essential parts:
the *foot*: the locomotive organ
the *head*: with eyes usually on antennae
the *visceral mass*: containing the digestive, excretory, circulatory and reproductive organs
the *mantle*: an epidermic wrap which secretes the shell
the *shell*.

THE VITAL FUNCTIONS

Nutrition is ensured by the digestive system, which consists of the *mouth*, placed on the lower part of the head and furnished with *jaws*. Slightly to the back of the mouth is the pharynx which contains the *radula* and the tongue. The *radula* is a ribbon-like organ bearing rows of transversal, tooth-like structures, the number and shape of which are characteristic of each genus and even each species. The salivary glands, which sometimes double as poisonous glands, emerge into the pharynx. The oesophagus leads to the stomach where food particles are partly digested; digestion is completed in the hepatopancreas. The intestine opens via the anus into the *mantle cavity*.

Circulation is carried out by the circulatory system. The heart pumps the blood (or haemolymph), oxygenated by the gills, towards the various organs. Apart from the iron-rich haemoglobin, haemolymph also conveys a second respiratory substance: *haemocyanine*, which contains copper.

The excretory system allows the animal to eliminate the refuse of nutrition, basically nitrate derivatives. The organs which ensure this function vary according to the various groups. One of the excretory canals is often transformed into a genital canal. Excretory canals emerge into the mantle cavity.

Respiration is carried out via a pair of gills situated in the *mantle cavity*.

The nervous system consists essentially of a pair of cerebral ganglions from which stem several nerve cords connected to the foot and the various organs. Sensitive organs are considerably diversified; they include photosensitive organs which serve as eyes; balance organs, chemically receptive organs sensitive to the various chemical substances; and tactile organs, placed on the mantle.

Reproduction in molluscs varies considerably. They are either gonochoric or hermaphroditic, which means that the sexes are either separate or within the same individual. They always have a reproductive organ (gonad) which produces the reproductive cells – sperms in the male and ova in the female. If the fertilisation takes place outside the animal, the cells are dispersed in the water; if inside, the male introduces the sperm via the duct to the genital duct of the female and in these cases a protective membrane is formed around the fertilised eggs.

The egg gives birth to a larval form or trochophore, which breaks out of the protective membrane and is transported over long distances to ensure the dispersion of the species. Usually the trochophore goes through a second stage as a veliger, a larva which already shows some of the characteristics of the adult in having foot, mantle, head, vibrating cilia designed to ensure locomotion and nutrition. Some of these veligers can be transported from one side of the oceans to the other.

MOLLUSCS AND THE ANIMAL KINGDOM

Classification:

Gastropoda: over 80,000 species characterised by:
 well-developed foot, sometimes provided with an *operculum*, either calcareous or chitinous
 visceral mass and shell, both coiled

Bivalvia (or lamellibranchia, or pelecypoda): about 20,000 species characterised by:
 a ventral, axe-shaped, burrowing foot
 visceral mass with lamellar gill
 a bivalve shell

Scaphopoda: about 1000 species characterised by:
 highly reduced head
 well-developed foot
 a well-developed visceral mass
 a shell shaped like an elephant's tusk, open at the two extremities – the larger opening at the front end is used by the foot, the one at the rear end ensures excretion and reproduction

Cephalopoda: about 700 species (plus a large number of fossil species) characterised by:
 a foot at the fore front, with arms covered in suckers
 well-developed head
 shell: non-existent (squid)
   internal (cuttle-fish)
   external with chambers (nautilus)
   external, secreted by two flattened arms of the female to contain the eggs (paper nautilus or argonaut)

Polyplacophora: (or chitons) about 500 species characterised by:
 a skeleton made up of eight calcareous layers

THE SHELL

The shell is formed by the mantle which secretes *conchiolin*, its proteinic support. It consists mainly of calcium carbonate ($CaCO_3$) based on the ions Ca^{++} supplied by water or soil. Calcium carbonate crystallises either as calcite or as aragonite. The shell reflects the mollusc's organisation and life, like a curriculum vitae; a detailed examination of it allows us to draw certain conclusions as to the age (juvenile, adult), the habitat (calm: rough shell; stormy: smooth shell), the accidents (scars, abnormal growths), ailments (malformations, spots, abnormal colourings), predators (perforations caused by naticas and other carnivorous molluscs). The shell can consist of three kinds of layers:

Periostracum: an organic membrane, usually brown, varying in thickness and ease of removal. This layer does not exist in such species as the cowries and others whose shell is extremely smooth.

Ostracum: a tough and thick layer with a crystalline structure.

Hypostracum: also called nacre, is thinner and is lamellar in structure.

The shapes and sculptures of shells are extremely variable: the pictograms give an idea of the main shapes and patterns. Certain cells contained in the mantle secrete pigments which are deposited in the ostracum according to genetic codes. These pigments are still imperfectly understood, which means that as yet no one has been able to supply a valid explanation for such aberrations as melanism and albinism.

Growth is rapid in the young individual, but slows down as soon as sexual maturity has been reached. Little is known about the growth and life spans of molluscs (except for those which are being bred, such as mussels and oysters). The methods used to mark the various populations have not yet yielded any definite results. Any collector of cowries can compare the adult specimens of any one species and note the considerable variations in size (up to three times).

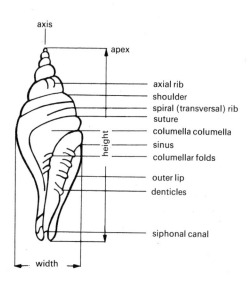

Figure 1

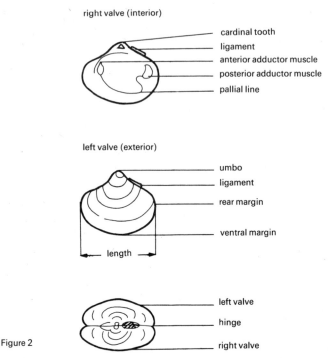

right valve (interior)

- cardinal tooth
- ligament
- anterior adductor muscle
- posterior adductor muscle
- pallial line

left valve (exterior)

- umbo
- ligament
- rear margin
- ventral margin

length

- left valve
- hinge
- right valve

Figure 2

The great majority of gastropods have *dextral* shells, coiling from left to right: the aperture, when facing the observer, is to the right of the columellar axis. When the opposite happens, the coiling is called *sinistral* (see the sacred chank of the Hindus *Turbinella pyrum* (Plate 88).

The *operculum* is a horny (thin and brownish) or calcareous (thick, white or coloured) structure which is used to close the aperture. The operculum is absent in certain families and/or in certain genera.

Figure 1 shows the shell of a gastropod, highly stylised. Figure 2 gives a schematic view of a bivalve.

THE MOLLUSCS' HABITAT

Molluscs are spread over all habitats of the globe: lowlands, forests, deserts, lakes, rivers, streams and oceans. They first appeared on earth during the Palaeozoic Period, that is about 500,000,000 years ago, and have since slowly evolved in the course of the various geological eras, adapting themselves to all sorts of biotopes.

Marine molluscs are said to be *benthic* if they live, fixed or mobile, on the substratum and *pelagic* if they move freely in the water. *Free benthic molluscs* have a well-developed foot and wander around looking for food – cones, spider conchs, bursae; others prefer to live in one place – limpets and

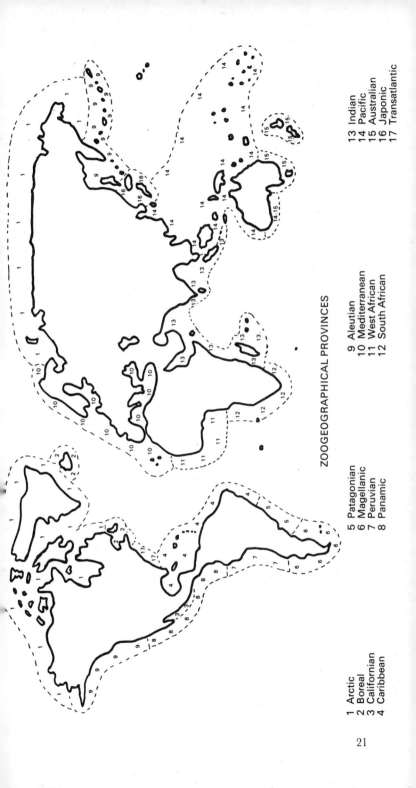

ZOOGEOGRAPHICAL PROVINCES

1 Arctic
2 Boreal
3 Californian
4 Caribbean

5 Patagonian
6 Magellanic
7 Peruvian
8 Panamic

9 Aleutian
10 Mediterranean
11 West African
12 South African

13 Indian
14 Pacific
15 Australian
16 Japonic
17 Transatlantic

21

chitons. Molluscs living on a shifting substratum (numerous gastropods such as olives and cones) often move under the sand itself, leaving their syphon to protrude and gather information on their surroundings. *Fixed benthic molluscs* (such as several bivalves) hang on to the substratum by means of very strong filaments called *byssus*, which are secreted by their foot (*Pinctada margaritifera, Tridacna maxima*); others weld one of their valves to their support (oysters, *Spondylus*). Types of anchorage are also associated with feeding on rocks or coral (*Lithophaga*) or wood (tarets).

Pelagic molluscs are either *planctonic* or *nectonic*. The former do not really swim, but let the currents move them around. The latter, a stage further in evolution (cephalopods like squids and octopuses) are good swimmers; they move around by ejecting the water collected within their mantle cavity. Certain bivalves, like the pectens, actually move by violently closing their valves.

The benthic world is divided into a *littoral zone*, reaching as far down as the areas where photosynthesis is still possible, and a *deep zone*. The littoral zone includes various divisions:

The *splash zone* – that part of the coast lying above the high-tide mark and almost never touched by water.

The *middle zone* – lying between the level reached by waves at high tide and the mark of the lowest tide (also called *intertidal zone*); this is where mangroves grow.

The *sub-littoral zone* – lower than the previous ones, this area can only be explored by skin-diving. It is here that madreporic and reef formations develop, the formations which are typical of intertropical regions between the latitudes 35°N 25°S, particularly in the huge Indian and Pacific areas, two of the provinces where the collecting and study of shells can best be carried out.

In the mid-nineteenth century, Samuel Pickworth Woodward studied the molluscan distribution in our seas and oceans and drew a map of seventeen zoogeographical provinces which is still used to this day with minor modifications (page 21).

ECOLOGY

In the twentieth century man is becoming increasingly conscious of the various links between living creatures and their habitats, and of how much the survival of the species depends on the environmental balance. However, marine ecology research is still very new indeed and its results incomplete. Apart from the geographical distribution of species which we have already mentioned, ecological research has concentrated on the quantitative distribution of species within certain sectors such as the water's edge, reefs, etc.

The quantity of molluscs can be expressed as an average per square metre: net weight and biomass (weight of molluscs less weight of shells). Corals, echinoderms, fish and seaweeds are recorded at the same time. Equally important is the prevalence or percentage of a certain species in relation to others within the area being examined. These measurements indicate the relationship between the biotope and the feeding habits of any given species. For example, at the water's edge the dominant species feed on seaweeds and other organic detritus, while carnivorous species are rarely found.

By carrying out these measurements over a period of time the influence of habitat on growth can be determined, as well as on the preservation and the renewal of the species. This kind of research is particularly valuable and easy in the case of molluscs, as they never move very far and live in very specific habitats. It is therefore obvious that a serious collector, respectful of life and habitat, has to follow certain rules:

1. He should collect as few living molluscs as possible. Any examination meant to lead to a wise selection but not carried out *in situ* only causes superfluous specimens to be abandoned on the beach or in a dustbin.

2. He should always have at hand the necessary equipment to keep the collected specimens in the best possible conditions.

3. He should respect the habitat. While the species living in shifting substrata can be collected easily, this is not the case for those living within corals. In the case of loose blocks or rocks, these should be carefully replaced after examination as they usually harbour eggs and a large number of living forms which are an important part of the ecological balance of the area.

Obviously the best solution is to photograph the animal in its biotope, and the collector can derive much pleasure from this practice. We should again stress the fact that to find (not as easy as it seems), photograph and collect a certain specimen, however common, will add great value to a collection, if not from the commercial angle, from the point of view of general interest because of the details and memories connected with it. Any information regarding the habitat will be much more precise and exact than in the case of a specimen acquired second or third hand.

Quite apart from the individual's self-control, hopefully each country protects its marine fauna with wise legislation aimed at curbing massive collecting on a commercial scale, often carried out without either knowledge or scruples and leading to irreversible damage to the environment. The institution of marine reserves, like those which already exist in many countries (Kenya, Israel, Australia) is an excellent solution and should be adopted by all countries wishing to preserve their marine heritage.

Pollution due to ecological imbalance: the intensive deforestation of certain tropical regions can erode soil which, once suspended in water, inevitably causes the death of coral reefs by asphyxiation.

The disposal of waste waters: rich as they are in organic matter, waste waters cause an unruly proliferation of algae. The environment becomes thus depleted of oxygen and unsuited to reef life. Besides, these domestic effluents carry a large number of pathogenic bacteria which can contaminate those bivalves which live by filtering water.

Thermal pollution: the spillage of high-temperature waters, often originating from power stations, can be catastrophic. The majority of corals can grow successfully only within a few degrees' variation – from 22°–28°C.

Chemical pollution: hydrocarbons floating on the surface of the water can be deposited on exposed reefs which would inevitably cause the death of corals and all small living organisms which inhabit them. Organo-chemical pesticides are particularly dangerous; once absorbed, they are transmitted from animal to animal. Certain effluents contain metallic residues of mercury, cadmium or lead which build up within the tissues of molluscs and endanger the consumer.

Nuclear pollution: nuclear explosions contaminate the environment in various degrees according to the frequency, power, altitude or depth of the experiments. Much more serious is the difficult problem of the disposal, at great depth, of thousands of tons each year of highly radioactive products having a very long period of lethal activity.

Visual pollution: very often the seabed is regarded, both by locals and by tourists, as a sort of dump. We have frequently found it impossible to photograph molluscs in their own environment because the latter was defaced by empty bottles, non-perishable plastic containers and lemonade or beer cans.

Illegal collecting and bad collecting methods: the seabed of some Mediterranean areas has been ravaged by people fishing for red coral (*Coralium rubrum*) simply by dragging with a rope-bound metal cross which breaks and tears coral branches, causing great damage for a small return. In the same areas fishermen using dynamite have, over the years, caused irreparable damage; for the few dead fish floating on the surface, how many sink to the bottom, their swimming bladder lacerated, and how many more animals are obliterated over a wide area? This is particularly regrettable because the study of corals and how they grow in a non-contaminated environment can result in very precise information as to the chronological evolution of atolls and coral reefs. Such an attitude is obviously due to ignorance which it is the duty of all collectors to remedy.

SHELL COLLECTING

Those shells one can find on a beach are usually dead and rarely intact: they have been battered by the waves and discoloured by the sun. However, they still provide valuable information of the frequency – or rarity – of a given species in a given environment. After a storm it is sometimes possible to find extremely rare specimens, which, although dead, have considerable value for the collector – for example the *Conus milneedwardsi*, on the coasts of Mozambique.

It is, however, advisable to look for live specimens. The majority of molluscs live hidden away: they avoid the light of day and use various marine formations as camouflage. One has to be a keen observer, particularly as a beginner, to find them. If the seabed is sandy, gastropods usually leave tracks in the form of furrows or swellings characteristic of each species; the animal will be found hidden at the end of such tracks. The mounds of seaweeds discarded by the waves or abandoned by the receding tide can yield innumerable small species: tribias, phasianellas, neritas, janthinas and spirulas.

Olives can be found at ebbtide, just under the sand soaked by the waves, preferably very early in the morning or at dusk: at the end of each track a lump of sand will hide the animal. At times, however, olives move about freely on the surface, and one can pick and choose at will. In Kenya, the natives gather them by the sackful to make soup. The more dirty the beach, the more this method of collecting is rewarding, particularly near restaurants or near fishing villages where food refuse is discarded as olives are carnivorous. Those dream beaches with white sands sparkling in the sun are barren places indeed!

Having discovered the tracks typical of members of the terebra and mitra families in a depth of 4–5 metres, one can collect them by floating vertically in the water and vigorously fanning one's flippers over the area; this will remove the superficial layer of sand which hides the molluscs, and they can then be easily selected and picked up by diving.

Large expanses of seaweeds are the favourite haunt of herbivorous species such as strombuses, haliotis, fasciolarias and cypraeas, and also of their predators – naticas and cones.

Should the seabed be rocky, one should carefully examine crevices, turn stones, search under corals, near wrecks, on the gorgonias, inside sponges and under sea anemones. Best results will be obtained at dusk or better still at night. Skin-diving apparatus, complete with pressurised air, enables one to explore less exploited areas, but this is forbidden by some governments, quite independently of the reserves where any means of collecting are outlawed.

Shells living at great depths are gathered by dredges or trapped in fishing nets. Certain species, very rare on the market, have only been found within the stomach of large fishes and their habitat is still totally unknown; this is the case of the *Cypraea fultoni* Sowerby.

THE ESSENTIAL EQUIPMENT

—a sack or narrow-mesh net to carry the specimens collected
—jars to contain small specimens or very fragile shells
—thick-soled waterproof shoes or diving boots; never walk barefoot – think of those sharp corals and of sea urchins . . .
—thick gloves, of the diving or sailing kind
—a small knife, indispensable for detaching limpets, chitons, etc
—small tweezers to extract shells from holes
—some protection against the sun (a *cotton* shirt, etc), at least in tropical countries
—should one wish to spend several hours in the water, it is important to consider the loss of body warmth, even in tropical areas where water temperature is seldom higher than 26° or 27°C near the surface, and be sensibly equipped.

CLEANING AND PRESERVING METHODS.

The value of a shell depends largely on the methods used to clean and preserve it. There are two ways to extract the animal from its shell:
Freezing: Place the specimen in a plastic bag and leave it in the deep freeze overnight. Leave it then to thaw out to normal temperature. This operation may have to be repeated two or three times. The adductor muscle which anchors the animal to its shell will thus be detached and it will be possible to extract the mollusc, slowly in order not to tear it. Should this happen, a jet of water will clear the last pieces out of the shell.
Boiling: Immerse the animal in sea water and slowly bring to the boil.

Allow to simmer for three to five minutes according to the size, then allow to cool. It will then be possible to extract the animal as above while it is still warm; avoid all sharp changes in temperature. This method is not advisable in the case of shells with large callosities as the grain of the thick parts could crack. However, we have often found (and photographed) living shells with naturally cracked callosities, even before any treatment had taken place.

If the specimens collected are relatively small, the simplest and quickest method is to use alcohol (it is however to be avoided with certain cypraeas, particularly the very dark ones, as irisation could result). The specimen should be placed in a small glass or plastic container and covered with a 70% solution of alcohol; after a few days, it should be removed from the solution and allowed to dry.

If the animal is not to be the object of study, the same method can be adopted by using concentrated formalin, but the same restrictions apply in the case of shiny and dark shells such as cypraeas and olives. More important specimens should be treated as follows. A few cubic centimetres of formalin should be injected with a syringe with an extremely fine needle, the total quantity being equally divided into injections in different spots. The author himself has often successfully used this method, particularly for *Cypraea tigris*.

Calcareous encrustations can be eliminated by immersing the shell in a solution of sodium hypochloride; the same is applied to the *periostracum*, except that in some cases it is useful to retain the latter, at least for a few specimens (eg cymatiums).

If the mollusc is to be properly studied, it should be preserved in 70% ethanol or 50% isopropylic alcohol. Sometimes on collecting expeditions it is necessary to take a small supply of these products as one leaves the larger towns, particularly in countries like Sri Lanka and in all islands without important conurbations where one might waste much time looking for one's supplies.

For the journey home, shells should be properly packed in plastic bags or small boxes, or in cartons such as those used for milk or fruit juices. All details which will be useful and even essential later on should be noted immediately. Certain shells can become friable as they dry out, and should therefore be soaked in a mixture of alcohol, glycerine and water in the proportions 2:2:6.

LABELLING

Once the shells are ready to be displayed, they should be accompanied by a label indicating the date of collection (or purchase), the place where found, the depth, the nature of the seabed, and so on. All these details are very important and greatly contribute to the value of the collection. On the label should also be noted the measurements: length, width and height (or the widest diameter for the Trochidae for instance), all indicated to 0·1 of a millimetre in the case of medium to small specimens. All labels should be written in Indian ink to prevent them from fading. Labels could be replaced by a number or a combination of numbers and letters referring to a register or card index. A letter will indicate the family

(V = Volutidae, M = Muricidae, etc) amd the numbers should indicate the chronological order in which the specimen has been added to the collection as a whole and to the family in particular.

Whatever the system used, it is always a good idea to keep a record showing not only the origin of the shell (country, habitat, etc) but also the intermediary who supplied it: dealer, collector, correspondent. It would also be extremely useful to mention – on label, card or register – the bibliographical sources upon which the identification of the shell was based. This information can also be supplied in coded form with letters or numbers referring to titles and authors. There are as many classification systems as there are collectors, and all have their good points and their shortcomings.

ARRANGING SHELLS

The beginner will probably find that cardboard boxes are adequate to contain his first discoveries or acquisitions. These boxes could be subdivided by smaller ones (all collectors' clubs can supply boxes, which can be folded flat, to provide a very practical modular system) or by cardboard walls or thin pieces of wood. The advantage of this method is that these first specimens will thus be sheltered from both dust and sunlight, which can fade certain colours within a few years.

However, the collector will soon realise that, even when the boxes are arranged within closed chests or sets of shelves, he will have to remove a whole pile of them in order to reach the one he wants – invariably the one at the bottom. There is only one foolproof system to avoid this: the right chest-of-drawers, which can be bought in either wood, metal or plastic, or can be made at home with some patience and a few tools.

There are also pieces of furniture usually made for offices offering horizontal classification. They are an ideal solution as they can contain up to 180 drawers of three different depths and are totally interchange-able. Although they represent a high initial expenditure, in the long term they will prove to give better value than a haphazard system which will never offer the same capacity in relation to a minimal use of space (40cm deep × 90cm wide × 180cm high). The authors themselves have adopted this system of cabinets and for years now have found it rewarding: all their shells as well as their slides are neatly arranged within the drawers. Should a collector also have at his disposal one or more glass-fronted cabinets to display his most beautiful specimens or his favourite family, he will then be a fortunate person.

A good solution, particularly where space is at a premium, is to ring the changes between, say, two or three families or the specimens collected during a recent journey. One would thus avoid exposing the same shells to the light for too long a time. In order to keep one's shells in the best possible conditions (this also applies to valuable, or indeed irreplaceable books), direct sunlight, too strong or prolonged artificial light, tobacco smoke, oak furniture (oak contains tannic acid which can affect the brilliant colours of certain shells like olives and cypraeas) and plastic foams containing chlorine should be avoided. If considered desirable, a little silicon oil can be used to heighten the brilliancy of a specimen or underline certain colours; the surplus oil can be removed with a soft cloth.

LIST OF ILLUSTRATED SPECIES

We have started, like most books dealing with molluscs, with the shells of gastropods which are the oldest and most primitive; these are followed by the families or genera derived from them.

We then go on to the bivalves and end with the most evolved species, the cephalopods. A few terrestrial gastropods, chosen from among the most spectacular ones, have been added over and above this classification.

The number of plates devoted to each family is not proportional to the importance of the family itself, although the relative importance of shell families as far as collectors are concerned has been kept in mind.

Class of Gastropoda (or univalves)

Family	Plate	Species (and occasionally sub-species)
Pleurotomariidae	1	*Pleurotomaria teramachii* Kuroda (*Perotrochus teramachii*)
Haliotidae	2	*Haliotis rufescens* Swainson
	3	*Haliotis scalaris* Leach
Fissurellidae	4	*Amblychilepas scutellum* Gmelin
Patellidae	5	*Patella longicostata* Lamarck
Trochidae	6	*Trochus niloticus* Linnaeus
	7	*Cittarium pica* Linnaeus
	8	*Clanculus pharaonius* Linnaeus
	9	*Maurea tigris* Martyn
	10	*Bathybembix argenteonitens* Lischke
	11	*Angaria melanacantha* Reeve
Turbinidae	12	*Turbo marmoratus* Linnaeus
	13	*Turbos divers* (opercula)
	14	*Turbo petholatus* Linnaeus
	15	*Phasianella* species
	16	*Astraea heliotropium* Martyn
	17	*Guildfordia yoka* Jousseaume
Neritidae	18	*Neritina communis* Quoy & Gaimard
Turitellidae	19	*Turritella terebra* Linnaeus
Architectonicidae	20	*Architectonica nobilis* Röding
Siliquariidae	21	*Siliquaria anguina* Linnaeus
Epitoniidae	22	*Epitonium scalare* Linnaeus *Epitonium pallasi* Kiener
	23	*Amaea magnifica* Sowerby
Strombidae	24	*Strombus aurisdianae* Linnaeus
	25	*Strombus canarium* Linnaeus
	26	*Strombus gallus* Linnaeus
	27	*Strombus gigas* Linnaeus

	28	*Strombus lentiginosus* Linnaeus
	29	*Strombus listeri* Gray
	30	*Strombus sinuatus* Humphrey
	31	*Lambis scorpius* Linnaeus
	32	*Lambis crocata* Link
	33	*Tibia fusus* Linnaeus
Xenophoridae	34	*Xenophora pallidula* Reeve
	35	*Stellaria solaris* Linnaeus
Naticidae	36	*Polinices albumen* Linnaeus
Cypraeidae	37	*Cypraea arabica*
		(*Mauritia arabica arabica* Linnaeus)
	38	*Cypraea argus*
		(*Talparia argus argus* Linnaeus)
	39	*Cypraea aurantium* Gmelin
	40	*Cypraea diluculum*
		(*Adusta diluculum* Reeve)
	41	*Cypraea errones*
		(*Erronea errones bimaculata* Gray)
	42	*Cypraea lynx*
		(*Cypraea lynx lynx* Linnaeus)
	43	*Cypraea mappa*
		(*Cypraea mappa mappa* Linnaeus)
	44	*Cypraea moneta*
		(*Monetaria moneta moneta* Linnaeus)
	45	*Cypraea ocellata*
		(*Erosaria ocellata* Linnaeus)
	46	*Cypraea tigris*
		(*Cypraea tigris tigris* Linnaeus)
Ovulidae	47	*Ovula ovum* Linnaeus
	48	*Calpurnus verrucosus* Linnaeus
	49	*Cyphoma gibbosa* Linnaeus
	50	*Cyphoma gibbosa* Linnaeus
		Cyphoma signata Pilsbry & MacGinty
		Cyphoma macgintyi Pilsbry
Cassididae	51	*Cypraecassis rufa* Linnaeus
	52	*Cassis tuberosa* Linnaeus
	53	*Casmaria ponderosa* Gmelin
	54	*Phalium strigatum* Gmelin
	55	*Morum cancellatum* Sowerby
Bursidae	56	*Bursa bubo* Linnaeus
	57	*Biplex perca* Perry
Cymatiidae	58	*Cymatium lotorium* Linnaeus
	59	*Charonia variegata* Lamarck
	60	*Distorsio anus* Linnaeus
Tonnidae	61	*Tonna perdix* Linnaeus
	62	*Malea ringens* Swainson
Muricidae	63	*Murex alabaster* Reeve
		(*Siratus alabaster*)
	64	*Murex haustellum*
		(*Haustellum haustellum* Linnaeus)
	65	*Murex orchidiflorus*

		(*Chicoreus subtilis* Houart 1977)
	66	*Murex palmarosae* Lamarck
	67	*Murex pecten* Lightfoot
	68	*Murex regius* Swainson
		(*Phyllotonus regius*)
Columbariidae	69	*Columbarium pagoda* Lesson
Coralliophylidae	70	*Rapa rapa* Linnaeus
	71	*Latiaxis pilsbryi* Hirase
	72	*Latiaxis divers*
Thaididae	73	*Thais planospira* Lamarck
	74	*Drupa grossularia* Röding
		Drupa morum Röding
		Drupa ricinus Linnaeus
Buccinidae	75	*Babylonia areolata* Link
		Babylonia spirata Linnaeus
	76	*Opeatostoma pseudodon* Burrow
Melongenidae	77	*Melongena corona* Gmelin
	78	*Syrinx aruanus* Linnaeus
	79	*Busycon contrarium* Conrad
Fasciolariidae	80	*Fasciolaria tulipa* Linnaeus
	81	*Fusinus colus* Linnaeus
Olividae	82	*Oliva oliva* Linnaeus
	83	*Oliva porphyria* Linnaeus
Mitridae	84	*Mitra mitra* Linnaeus
	85	*Mitra mitra* Linnaeus (cross-section)
	86	*Vexillum regina* Sowerby
		Vexillum taeniatum Lamarck
		Vexillum filiareginae Cate
Vasidae	87	*Vasum tubiferum* Anton
Turbinellidae	88	*Turbinella pyrum* Linnaeus
Harpidae	89	*Harpa harpa* Linnaeus
	90	*Harpa doris* Röding
Volutidae	91	*Amoria damonii* Gray
	92	*Lyria delessertiana* Petit de la Saussaie
	93	*Voluta ebraea* Linnaeus
	94	*Voluta imperialis*
		(*Cymbiola imperialis* Lightfoot)
	95	*Melo melo* Solander
	96	*Cymbiola nobilis* Lightfoot
	97	*Voluta nobilis*
		(*Cymbiola nobilis* Lightfoot)
	98	*Cymbiola vespertilio* Linnaeus
	99	*Cymbiola vespertilio* Linnaeus
Marginellidae	100	*Marginella angustata* Sowerby
	101	*Marginella cingulata* Dillwyn
		(*Persicula cingulata*)
		Marginella marginata Born
		(*Persicula marginata*)
	102	*Marginella pringlei*
		Marginella (Afrivoluta) pringlei Tomlin
Turridae	103	*Turris babylonia* (?) Linnaeus

	104	*Thatcheria mirabilis* Angas
Conidae	105	*Conus ammiralis* Linnaeus
	106	*Conus barthelemyi* Bernardi
	107	*Conus bullatus* Linnaeus (*Cylinder bullatus*)
	108	*Conus generalis* Linnaeus (*Leptoconus generalis*)
	109	*Conus genuanus* Linnaeus
	110	*Conus geographus* Linnaeus (*Gastridium geographus*)
	111	*Conus gloriamaris* Chemnitz (*Leptoconus gloriamaris*)
	112	*Conus praelatus* Hwass in Bruguiere
	113	*Conus pulcher* Lightfoot
	114	*Conus sozoni* Bartsch
	115	*Conus textile* Linnaeus (*Darioconus textile*)
Terebridae	116	*Terebra maculata* Linnaeus
	117	*Terebra triseriata* Gray
Hydatinidae	118	*Hydatina albocincta* van der Hoeven
	119	*Hydatina physis* Linnaeus

Class of Pelecypoda (or bivalves)

	120	*Pecten flabellum* Gmelin
Pectinidae	120	*Pecten flabellum* Gmelin
	121	*Pecten glaber* Linnaeus
	122	*Pecten speciosus* Reeve (*Chlamys speciosus*)
Spondylidae	123	*Spondylus americanus* Hermann
	124	*Spondylus regius* Linnaeus
Ostreidae	125	*Lopha cristagalli* Linnaeus
Chamidae	126	*Chama lazarus* Linnaeus
Cardiidae	127	*Corculum cardissa* Linnaeus
	128	*Cardium costatum* Linnaeus
Tridacnidae	129	*Tridacna gigas* Linnaeus
	130	*Hippopus hippopus* Linnaeus
Cutellidae	131	*Siliqua radiata* Linnaeus
Veneridae	132	*Pitar lupinaria* Lesson
	133	*Callanaitis disjecta* Perry
	134	*Venus lamellata* Schumacher
Glossidae	135	*Glossus moltkianus* Spengler
Pholadidae	136	*Barnea costata* Linnaeus

Class of Scaphopoda

Dentalidae	137	*Dentalium elephantinum* Linnaeus

Class of Cephalopoda

Argonautidae	138	*Argonauta argo* Linnaeus
Nautilidae	139	*Nautilius pompilius* Linnaeus
Spirulidae	140	*Spirula spirula* Linnaeus

Class of Polyplacophora (or chitons)

Land Shells

PLATES

Technical note

All the photographs in this book have been shot on Kodachrome 25ASA, the majority in daylight. When there was not enough light, we used a small flash placed at about 50cm from the subject, with an aperture of about f16. The camera used was a Leica R3, with a Macro-Elmarit 60mm lens, aperture f2.8. This limited equipment enabled us to take photographs, without any other accessory, from landscapes to objects smaller than 50mm (the smallest field being 50 × 70mm). With the extension tube supplied with this lens one can achieve a ratio of 1:1 (or a field of 24 × 36mm).

This camera, wrapped in a Turkish towel together with a few films, is always placed in a watertight sack specially made for boats. The equipment is thus perfectly protected from any impact, spray and particularly sand which can get everywhere. Underwater photographs were taken with the same Leica R3, a 35mm lens, aperture f2 and a minimum distance of 30cm, from within an underwater case we designed ourselves.

Perotrochus teramachii Kuroda   Pleurotomariidae
Teramachi's slit shell

A primitive gastropod which had long been considered extinct. Characterised by an anal fissure used for the excretion of refuse. Sixteen living species are known at present.

Dist: Caribbean and Japonic provinces   80–120mm

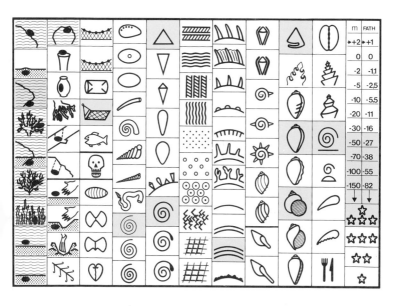

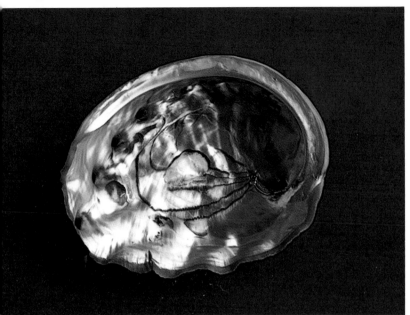

Haliotis rufescens Swainson 1822   Haliotidae
Red abalone

A shell with a reddish, very thick exterior. Considered edible in the USA and eaten as a steak. Controlled fishing.

Dist: Californian and Panamic provinces   → 300mm

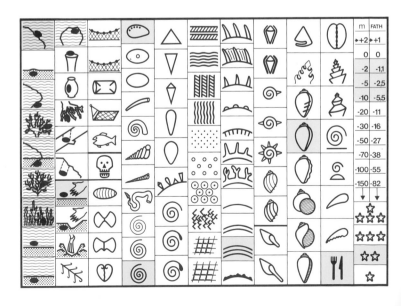

Haliotis scalaris Leach 1814
Syn: **H. tricostalis**

Haliotidae

Has 4–6 orifices through which water is expelled after oxygenating the gills.
Has a very large foot to anchor itself on rocks. The mantle covering the shell is
orange-brown. Vegetarian.

Dist: Australian province

80–100mm

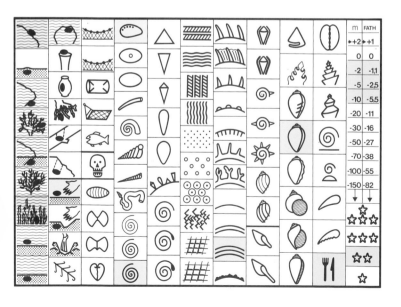

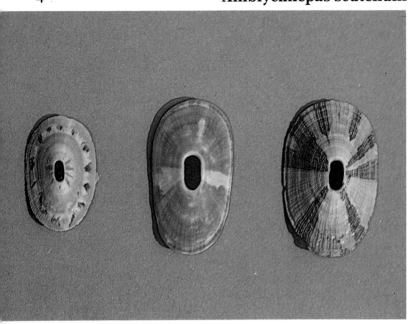

Amblychilepas scutellum Gmelin 1791   Fissurellidae

They live anchored to rocks in the tidal area by a sucker-like foot, in common with all fissurellids and patellids.

Dist: Australian province   25–30mm

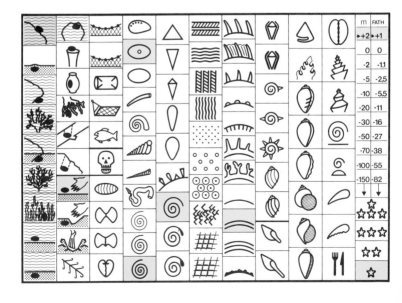

Patella longicostata Lamarck 1819   Patellidae

Vegetarian species. The shell is very dark on the outside.

Dist: South African province   30–80mm

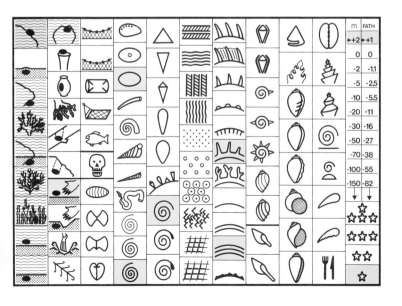

Trochus niloticus Linnaeus 1758   Trochidae
Commercial trochus, Nile trochus, Pearly-top shell

Trochuses have a thick shell lined with nacre on the inside. It is this nacre, or
mother-of-pearl, which makes them exploited for buttons, etc, particularly in
French Polynesia.

Dist: Indian and Pacific provinces   80–130mm

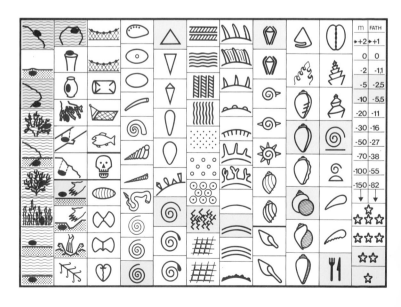

Cittarium pica Linnaeus 1758   Trochidae
West Indian top shell

Eaten as a delicacy in the Antilles.

Dist: Caribbean province   60–100mm

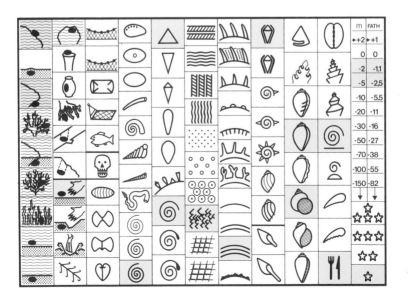

Clanculus pharaonius Linnaeus 1758   Trochidae
Strawberry top

The most beautiful representative of the genus **Clanculus** which includes about fifty species. It has a deep, cushion-shaped umbilicum.

Dist: Indian Ocean   15–20mm

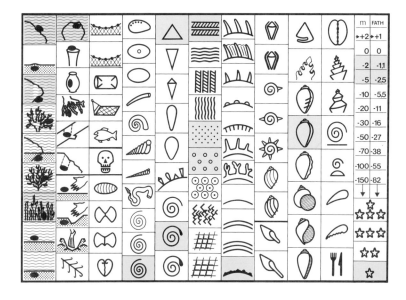

Maurea tigris Martyn 1794   Trochidae
Tiger top shell
Dist: Pacific province, New Zealand   30–60mm

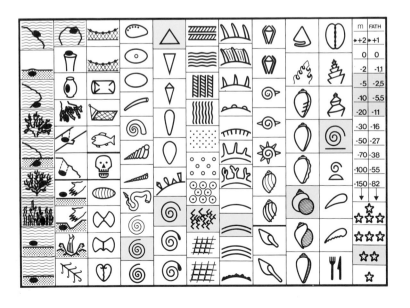

Bathybembix argenteonitens Lischke 1871   Trochidae

The genus **Bathybembix** includes ten species all living in very deep waters.

Dist: Japonic province   30–60mm

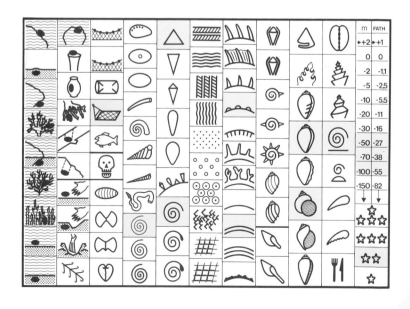

Angaria melanacantha Reeve 1842   Trochidae

A genus including a dozen species and once classed with the Turbinidae. All angarias have a horny operculum, like the Trochidae, and a nacre interior.

Dist: endemic to the Philippines   50–70mm

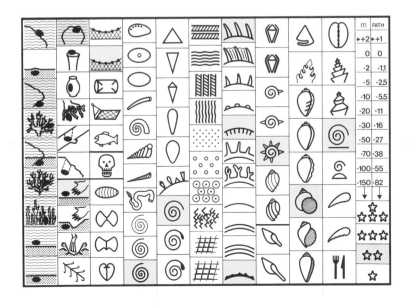

Turbo marmoratus Linnaeus 1758   Turbinidae
Green turban, Green snail

The largest of the turbans. Its very thick, nacred shell is used to make buttons.
Its calcareous operculum can weigh up to 500g.

Dist: throughout the Indian and Pacific provinces   → 200mm

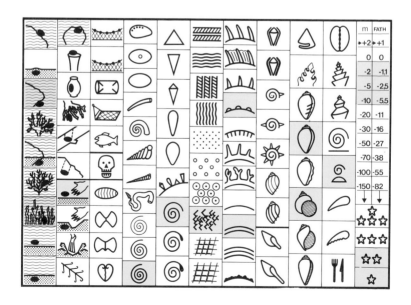

Turbans' opercula (various)   Turbinidae

The opercula of this family vie with the shells themselves in variety and beauty. From left to right and top to bottom: **T. petholatus**, **T. whitlyi**, **T. fluctuosus**, **T. sermaticus**.

Dist: Indian and Pacific provinces   12–35mm

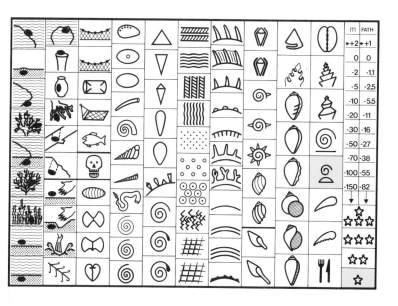

Turbo petholatus Linnaeus   Turbinidae
Tapestry turban

One of the most spectacular members of the large family of the Turbinidae which includes over 500 species. The beautiful calcareous operculum with its green and blue tints is known as 'cat's eye'.

Dist: throughout the Indian and Pacific provinces   40–80mm

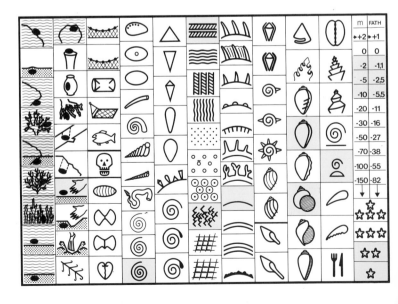

Phasianella species   Turbinidae
Pheasant shells

There are about forty species of pheasant shells, the largest of which, **P. australis**, can reach 80mm. The specimens illustrated here were collected at low tide, under dried seaweed, on the beach of Trincomalee (north-east Sri Lanka).   12–18mm

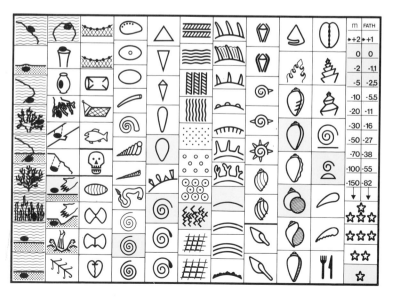

Astraea heliotropium Martyn 1784   Turbinidae
Sun shell, Sunburst star shell

The genus **Astraea** includes some fifty species, all conical but varying in ornamentation: their margin can be smooth, wavy or saw-toothed.

Dist: Australian province   80–100mm

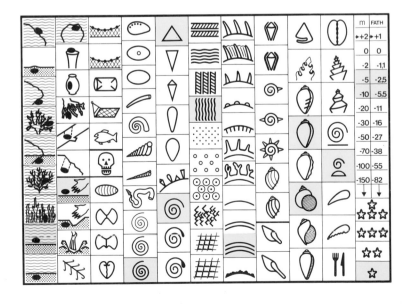

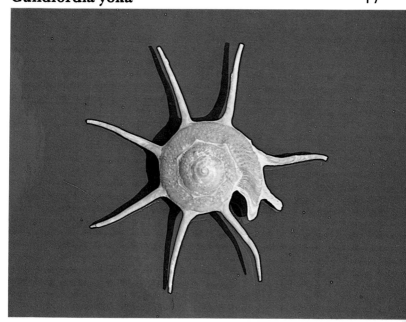

Guildfordia yoka Jousseaume   Turbinidae
Yoka star shell

The top spiral of the graceful shell is adorned with 7 to 9 digitiform spines.
Ovoid operculum.

Dist: Pacific and Japonic provinces   70–100mm

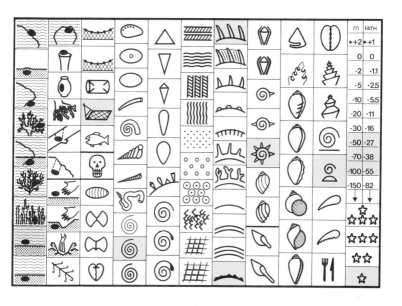

Neritina communis Quoy & Gaimard 1832   Neritidae
Syn **Theodoxus communis** or **T. pictoneritina**
Common nerite

Small shells with infinitely variable patterns and colours. Their habitats also
vary and include brackish waters.

Dist: Indian and Pacific provinces   10–15mm

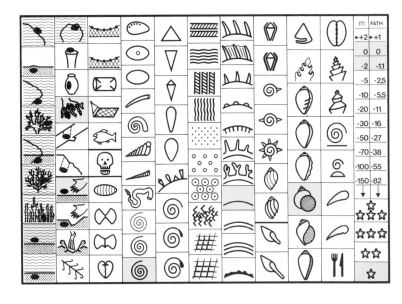

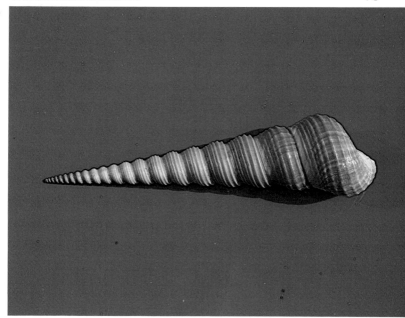

Turritella terebra Linnaeus 1758　　　　　　　Turitellidae
Auger screw shell

The Turritellidae include several hundreds of species living in almost all the seas.
They are carnivorous. **T. terebra** has over 25 coils.

Dist: Indian and Pacific provinces　　　　　　　100–180mm

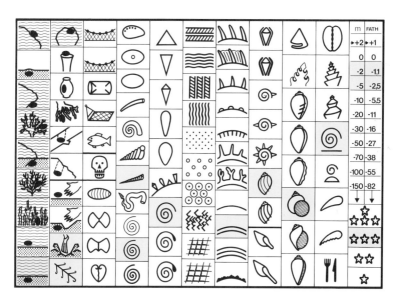

Architectonica nobilis Röding 1758   Architectonicidae
Sundial

This genus includes about forty species. The centre of this shell is deeply
scooped.

Dist: Caribbean province   25–50mm

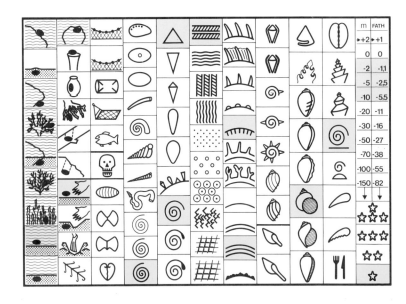

Siliquaria anguina Linnaeus 1758 Siliquariidae
Wormshell

The Siliquaridae feed on minute vegetable debris. They often form colonies.
The shells in this small family are characterised by the irregularity of their end
coils. Otherwise the coils are similar to those of Turbinidae.

Dist: throughout the Indian and Pacific provinces 60–120mm

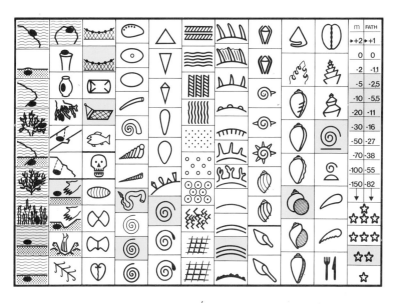

Epitonium scalare Linnaeus 1758   Epitoniidae
Epitonium pallasi Kiener 1838
Syn **Scalaria pretiosa** Lamarck 1816
Precious wentletrap

Characterised by the pronounced ribs which link the various coils. In the past it was highly appreciated by collectors (see Historical Introduction).

Dist: Indian and Pacific provinces   50–90mm

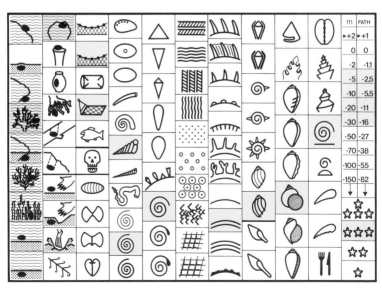

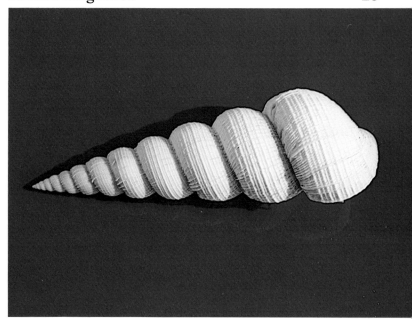

Amaea magnifica Sowerby 1844   Epitoniidae
Syn **Epitonium magnificum**
Magnificent wentletrap

One of the largest members of this family, it lives only at great depths. It is usually sold by fishermen in Taiwan.

Dist: Japonic province   80–120mm

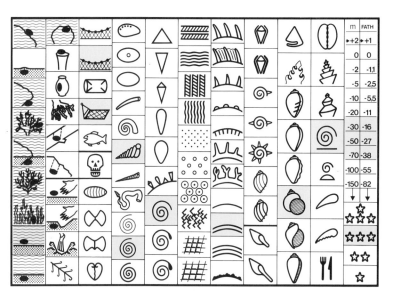

Strombus aurisdianae Linnaeus 1758   Strombidae
Diana's conch

There are about eighty species of Strombidae. They move quite rapidly by throwing forward their foot which is capped by an operculum shaped like a sickle, with which they anchor themselves. They are generally herbivorous.

Dist: throughout the Indian and Pacific provinces   50–70mm

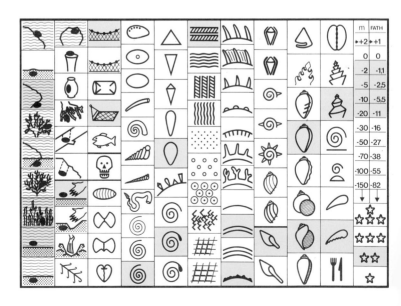

Strombus canarium Linnaeus 1758   Strombidae
Yellow conch, Dog conch
Syn **Lambis turturella** Röding 1798 and **S. isabella** Lamarck
1822

Colour extremely variable. Lives in colonies.

Dist: Indian and Pacific provinces   30–90mm

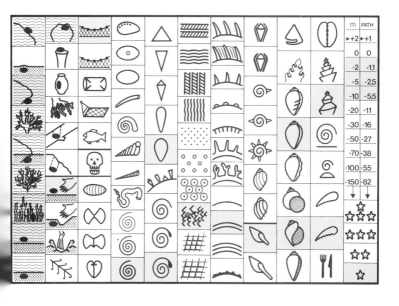

Strombus gallus Linnaeus 1758   Strombidae
Rooster-tail conch

A very beautiful conch, called **gallus** (cockerel) because of its 'wings'. Often found in the nets of the fishermen of Guadaloupe. Light brown periostracum, variable colour.

Dist: Caribbean province   120–160mm

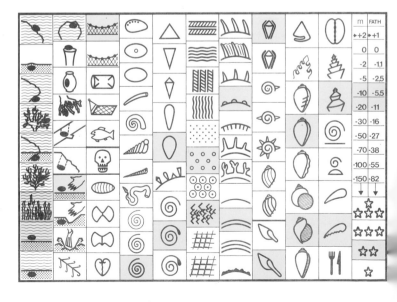

Strombus gigas Linnaeus 1758   Strombidae
Pink conch, Queen conch

It sometimes produces beautiful pink, semi-precious pearls. Used as an ornament as well as a kind of bugle. It is used to make delicious soups and escalopes.

Dist: Caribbean province   150–300mm

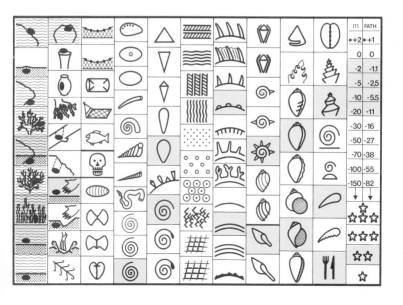

Strombus lentiginosus Linnaeus 1758   Strombidae
Silver conch

The coloured eyes at the tip of a peduncle are characteristic of each **Strombus**
species.

Dist: Indian and Pacific provinces   50–100mm

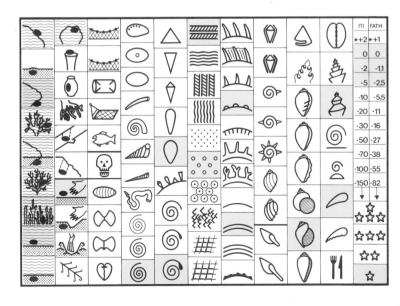

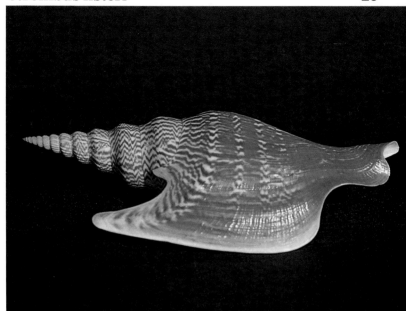

Strombus listeri Gray 1852   Strombidae
Lister's conch

This conch, today quite common, was considered extremely rare during the nineteenth century and the few known specimens fetched incredibly high prices.

Dist: Indian province   100–130mm

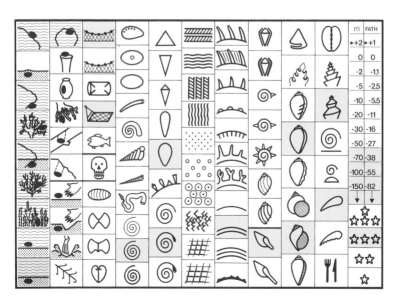

Strombus sinuatus Humphrey 1786 Strombidae

This beautiful stromb, not very common in the Indo-Pacific, has been found in large quantities at Bohol in the Philippines.

Dist: Indian Ocean 90–120mm

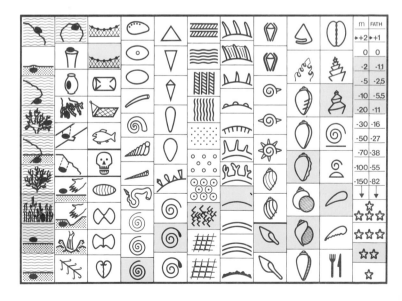

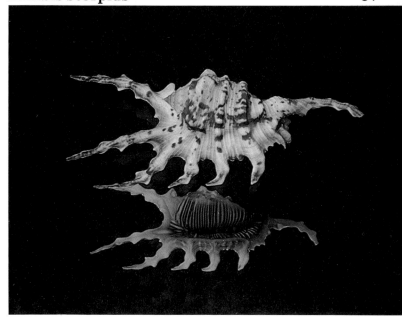

Lambis scorpius Linnaeus 1758   Strombidae

The molluscs of this genus are characterised by quite obvious sexual dimorphisms: males are smaller than females and the fifth digitation of the latter is straight.

Dist: throughout the Indian and Pacific provinces   100–150mm

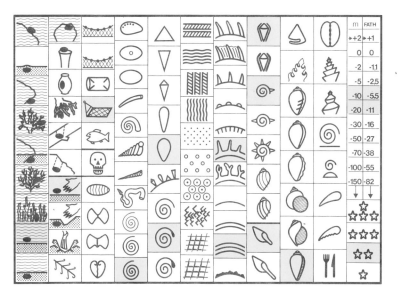

Lambis crocata Link 1807   Strombidae
Orange spider conch

A sub-species is also known, the **Lambis crocata pilsbryi** Abbott 1961,
which has longer digitations and is endemic to the Marquesas Islands.

Dist: throughout the Indian and Pacific provinces   100–150mm

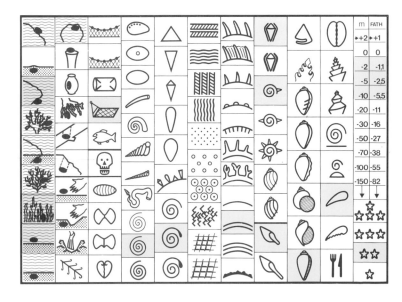

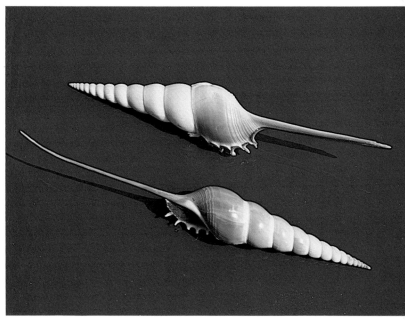

Tibia fusus Linnaeus 1758 — Strombidae
Spindle tibia

A very elegant shell, with the anterior syphonal canal as long as the shell itself.
Lives on muddy seabeds where visibility is very poor.

Dist: Indian and Pacific provinces — →300mm

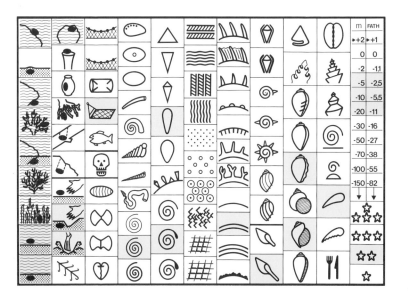

Xenophora pallidula Reeve 1842   Xenophoridae
Pallid carrier shell

Most members of this family have other shells, pieces of shells, or even bits of
rock attached to the exterior of their own shell.

Dist: Pacific and Japonic provinces   50–100mm

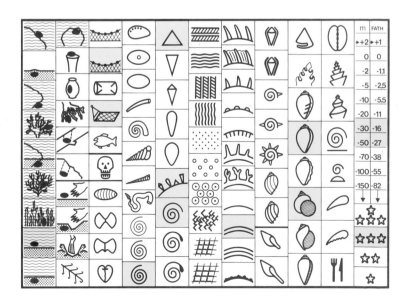

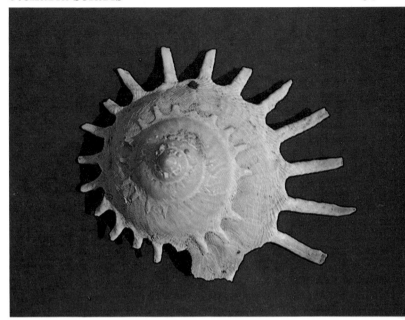

Stellaria solaris Linnaeus 1758   Xenophoridae
Sunburst carrier shell

The first specimen was found at the beginning of the eighteenth century. They were found in great numbers by Filipino trawlers during the Second World War.

Dist: South Africa, Indian and Pacific provinces   70–90mm

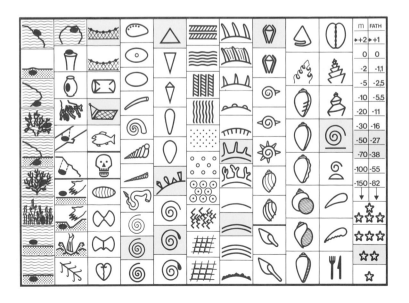

Polinices albumen Linnaeus 1758   Naticidae
Natice albumen

The naticids are carnivorous and perforate the shell of their victims to extract the flesh. Their eggs, laid in groups on the sand, form graceful, undulating chains. Their well-developed foot can cover the whole shell.

Dist: throughout the Indian and Pacific provinces   30–50mm

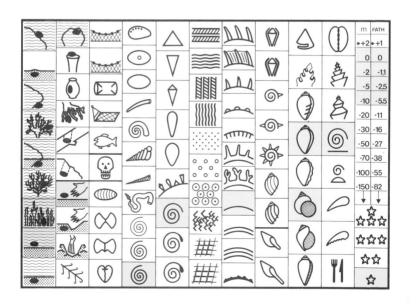

Mauritia arabica arabica Linnaeus   Cypraeidae
Arabian cowry

This genus includes five sub-species: **C.a. asiatica**, **C.a. dilacerata**, **C.a. immanis** and **C.a. westralis**, plus the one illustrated above; they all differ slightly in shape and pattern.

Dist: Indian and Pacific provinces   30–80mm

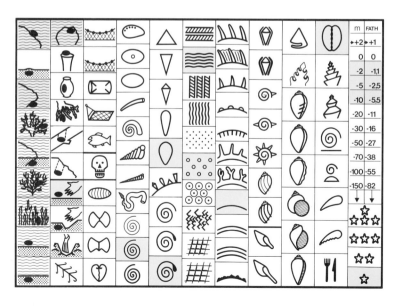

Talparia argus argus Linnaeus 1758   Cypraeidae
Eyed cowry

A cylindrical cyprea characterised by its ocelli (or eyes), which have given it its name (with reference to the Greek prince and his hundred eyes). Top: **C.a. ventricosa** Gray 1824; left: **C. a. argus** Linnaeus 1758; right: **C. a. constrastriata** Perry 1811.

Dist: Indian and Pacific provinces   65–95mm

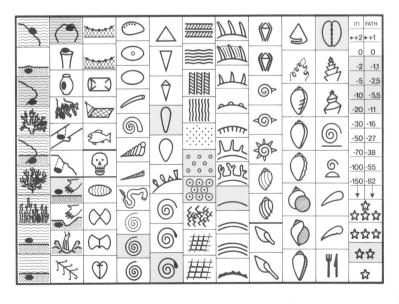

Cypraea aurantium Gmelin 1791   Cypraeidae
Golden cowry

Probably the best known and most sought after of all cypraeas by collectors. Its shape and its orange colour are unique.

Dist: Pacific province   60–110mm

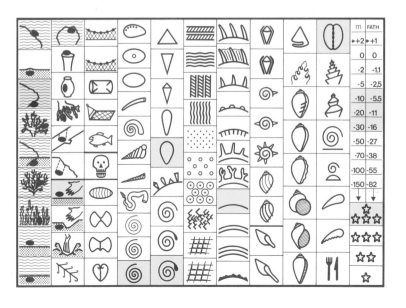

Adusta diluculum Reeve 1845   Cypraeidae

On the left is a photograph of **Palmadusta ziczac** Linnaeus 1758 (Zigzag cowry) which clearly shows the general differences between these two shells, both found in Kenya.

Dist: Indian province   *z.z.:* 13–20mm; *dil.:* 15–30mm

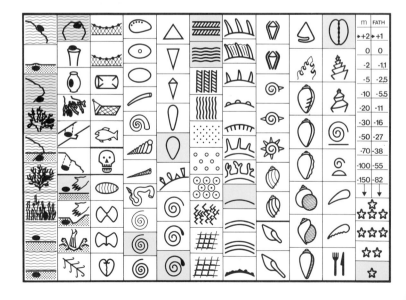

Erronea errones bimaculata Gray 1824   Cypraeidae
Wandering cowry

This photograph was taken *in situ* at Upuveli (Sri Lanka, near Trincomalee)
and shows a pair of cypraeas and their eggs on a rock covered with algae and
corals, at a depth of 2m in August 1975. Once photographed, they were wisely
left alone.

Dist: throughout the Indian and Pacific provinces   20–40mm

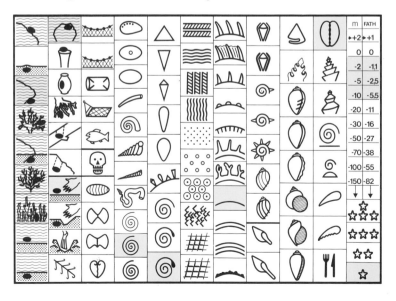

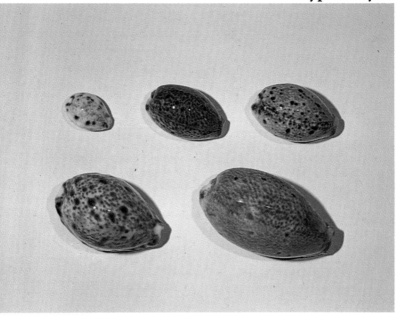

Cypraea lynx lynx Linnaeus 1758   Cypraeidae
Lynx cowry

This shell is very variable even within the type species. There are three sub-species, all located within the Indian and Pacific provinces.

Dist: throughout the Indian and Pacific provinces   20–70mm

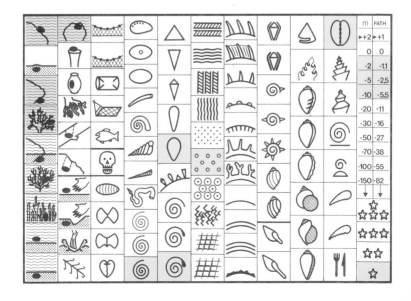

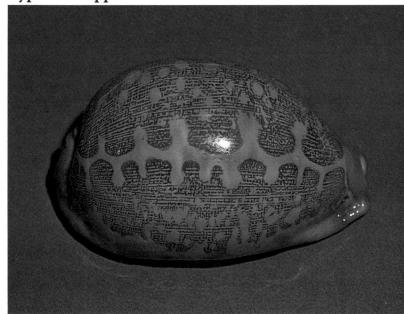

Cypraea mappa mappa Linnaeus 1758   Cypraeidae
Map cowry

This species has three sub-species: **C.m. alga**, **C.m. geographica** and **C.m. viridis**. The variations in shape and pattern which can be observed in the type species have caused certain authorities to question the sub-species. It is one of the most beautiful of cypraeas, but it is impossible to find two specimens exactly the same.

Dist: throughout the Indian and Pacific provinces   45–90mm

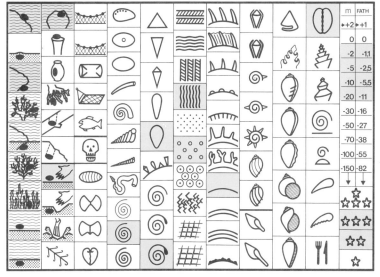

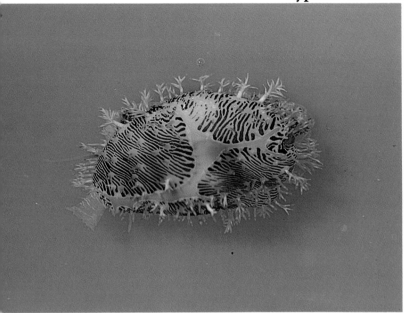

Monetaria moneta moneta Linnaeus 1758   Cypraeidae
Money cowry

Certain tribes have for centuries used this shell as currency and ornament for their costumes and hairstyles. The mollusc has a beautiful mantle with black stripes.

Dist: throughout the Indian and Pacific provinces   →40mm

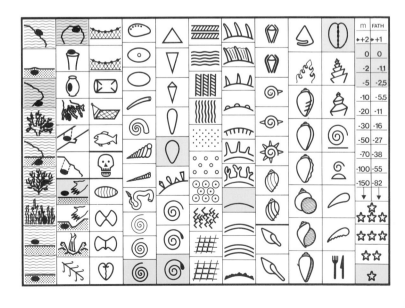

Erosaria ocellata Linnaeus 1758   Cypraeidae

This small cypraea varies very much in shape, but there are no sub-species.
Unique for its beautiful ocelli.

Dist: Indian province   12–30mm

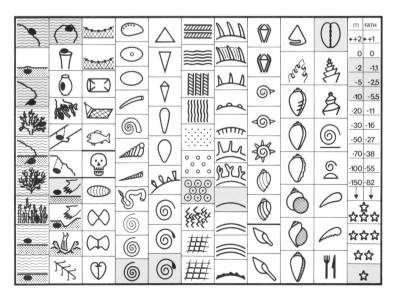

Cypraea tigris tigris Linnaeus 1758   Cypraeidae

Very variable in colour and size and including four sub-species, of which **C.t. schilderiana** can be up to 15cm long. The beautiful mantle is finely striped with brown.

Dist: throughout the Indian and Pacific provinces   40–150mm

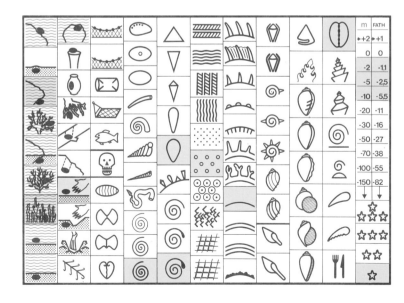

Ovula ovum Linnaeus 1758   Ovulidae
Egg cowry

This family includes over 200 members but many are too small to be of interest to the amateur collector. The shell is deep red on the inside. The black mantle can cover the whole shell, as with the cypraeas. It lives in association with certain celenterates upon which it feeds.

Dist: Indian and Pacific provinces   60–100mm

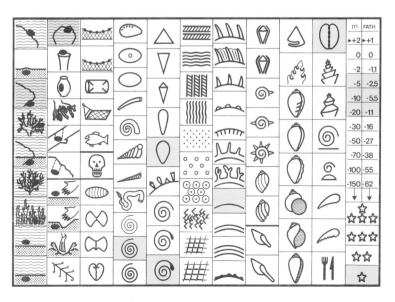

Calpurnus verrucosus Linnaeus 1758   Ovulidae

This mollusc is carnivorous, like all the ovulids. It clings to gorgonias and madrepores. The pustular mantle perfectly imitates the surface of the alcyonarians on which the animal lives.

Dist: throughout the Indian and Pacific provinces   20–30mm

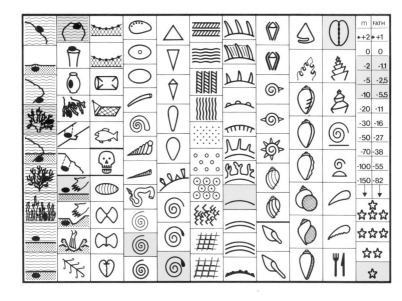

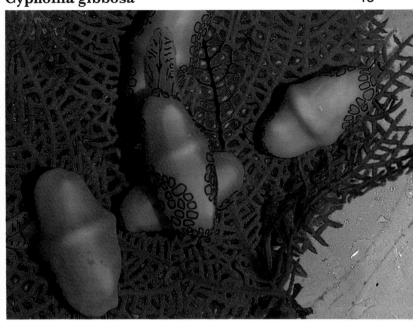

Cyphoma gibbosa Linnaeus 1758
Flamingo tongue

Ovulidae

This group of **C. gibbosa** was photographed by the authors in Martinique. They can be seen on a gorgonia, their favourite habitat. The beautiful pattern of the mantle, different in each species of **Cyphoma**, can be easily seen.

Dist: Caribbean province

15–40mm

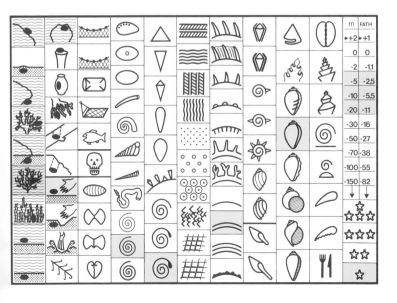

Cyphoma gibbosa Linnaeus 1758   Ovulidae
Flamingo tongue

The beautiful pattern of the mantle is different in each species. Left: **C. gibbosa**; centre: **C. signata** Pilsbry & MacGinty 1939; right: **C. macgintyi** Pilsbry 1939.

Dist: Caribbean province   20–35mm

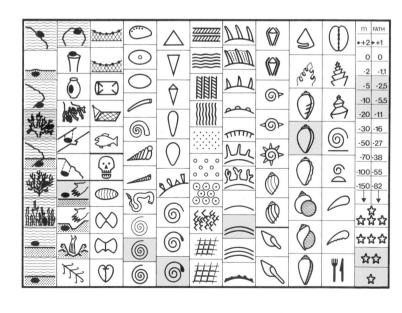

Cypraecassis rufa Linnaeus 1758   Cassididae
Bull's-mouth conch; Red helmet
The very thick shell is used in the manufacture of cameos.

Dist: Indian and Pacific provinces   100–180mm

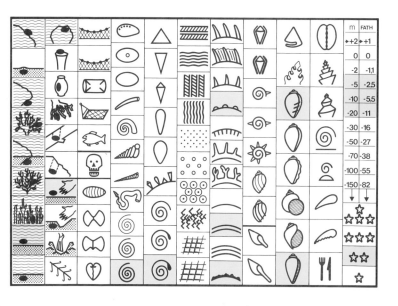

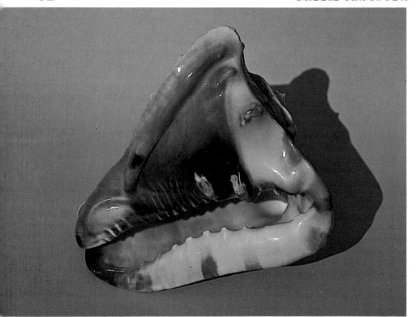

Cassis tuberosa Linnaeus 1758   Cassididae
King helmet

This family includes seven genera and eighty species, some of which are characterised by sexual dimorphism. The females of the illustrated species are larger than the males.

Dist: Caribbean province   →250mm

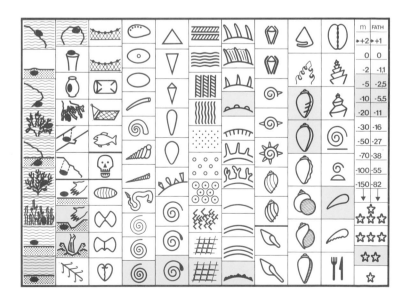

Casmaria ponderosa Gmelin 1791   Cassididae

The specimen on the left is 101mm long (probably a world record); it was bought by the authors in a souvenir shop in Hong Kong.

Dist: Indian and Pacific provinces   40–101mm

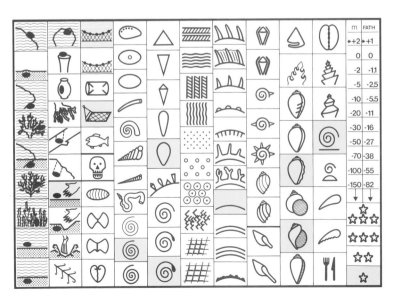

Phalium strigatum Gmelin 1791   Cassididae
Striped bonnet

A widespread genus with variable patterns. The external lip can sometimes have short digitations.

Dist: Indian and Pacific provinces   70–100mm

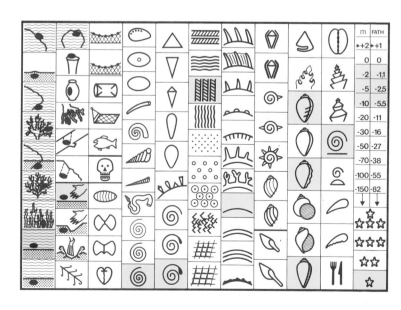

Morum cancellatum Sowerby 1824 Cassididae

The genus includes about twenty species, some of them very rare. The axial ribs intersected by the spiral ribs give it a characteristic pattern.

Dist: Indian and Pacific provinces 35–50mm

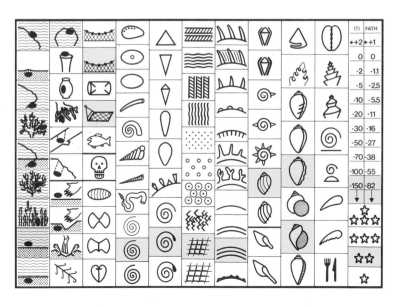

Bursa bubo Linnaeus 1758   Bursidae
Giant frog shell

This family includes over sixty species. The largest specimens were once used as oil lamps.

Dist: Pacific province   → 200mm

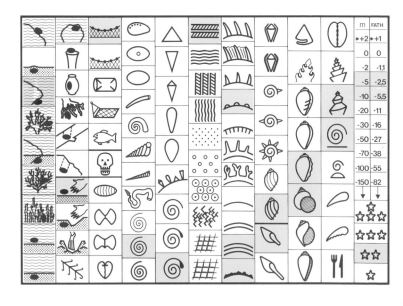

Biplex perca Perry 1811   Bursidae
Syn **Gyrineum perca**
Winged frog shell

Once classified among the Cymatidae. The shell is noticeably flattened – 8mm thick by 40mm wide. The genus **Biplex** includes half a dozen species.

Dist: Pacific and Japonic provinces   40–80mm

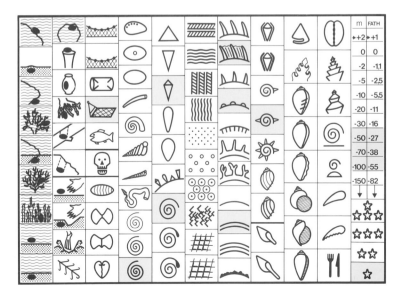

Cymatium lotorium Linnaeus 1767   Cymatiidae

A family of carnivorous molluscs whose secretion paralyses their victims. They can have hairs on their periostracum.

Dist: Indian, Pacific and Japonic provinces   100–160mm

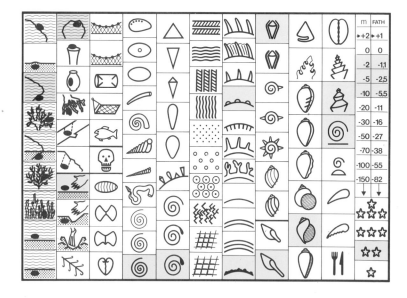

Charonia variegata Lamarck 1816   Cymatiidae
Atlantic triton

The larger specimens are used as trumpets by natives. It is now much less common than it used to be. The carnivorous charonias feed on acanthasters.

Dist: Caribbean province; Indian and Pacific provinces for **C. tritonis**

200–400mm

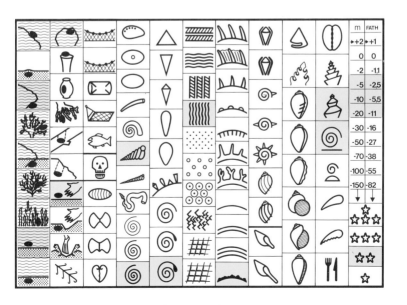

Distorsio anus Linnaeus 1767   Cymatiidae

The axis of each spiral is inclined at an angle to the preceding one, hence the name 'distorsio'.

Dist: Indian and Pacific provinces   40–80mm

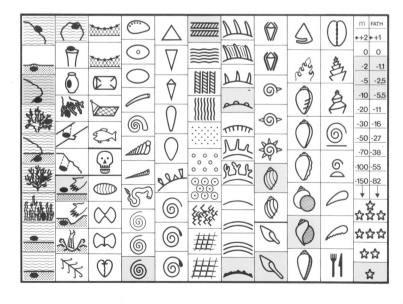

Tonna perdix Linnaeus 1758   Tonnidae
Syn **Dolium perdix**
Partridge tun

Its pattern and brown colour are similar to those of a partridge, hence its name;
the axial ribs are rounded.

Dist: Indian and Pacific provinces   → 150mm

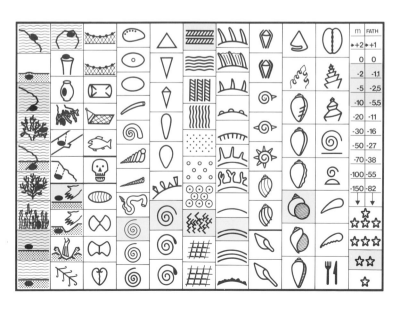

Malea ringens Swainson 1822   Tonnidae
Great grinning tun

The Tonnidae feed on echinoderms and crustaceans, whose carapaces they attack with their extremely acid and toxic secretions, dissolving the shell and paralysing the animal within. The marked axial ribs are rounded.

Dist: Panamic and Pacific provinces   120–170mm

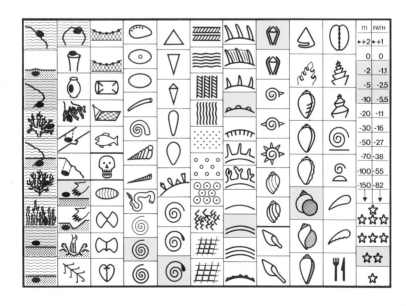

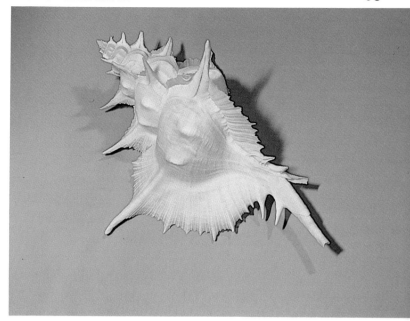

Siratus alabaster Reeve 1845
Abyssal murex

Muricidae

A very beautiful but fragile representative of the Muricidae family, which includes at least 400 species throughout the world, each varying considerably in shape and colour. They are carnivorous, and many are capable of drilling holes.

Dist: Pacific and Japonic provinces

100–150mm

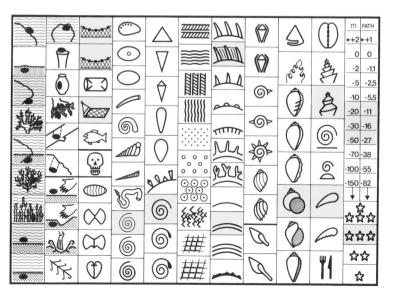

Haustellum haustellum Linnaeus 1758
Muricidae
Snipe's bill

A murex originating mainly in the waters of Japan, at least until a few years ago. Today beautiful specimens at affordable prices have been put on the market by the Filipinos.

Dist: Indian and Pacific provinces
70–180mm

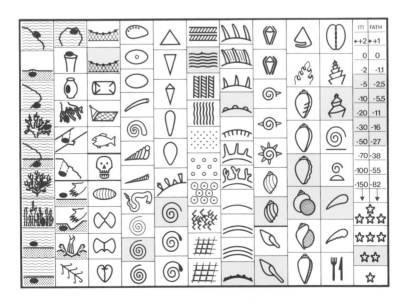

Chicoreus subtilis Houart 1977   Muricidae

A very beautiful, very fragile shell. It is gathered by Filipino fishermen with special nets which go down to about 100 fathoms (*c*180m).

Dist: Philippines   25–45mm

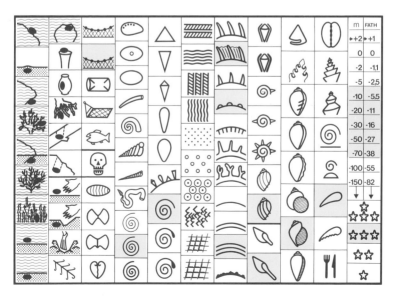

Murex palmarosae Lamarck 1822   Muricidae
Rose-branch murex

A splendid murex with wide, leaf-like digitations. It is very difficult to find a
perfect specimen, without encrustations of algae or corals.

Dist: Indian and Pacific provinces   70–130mm

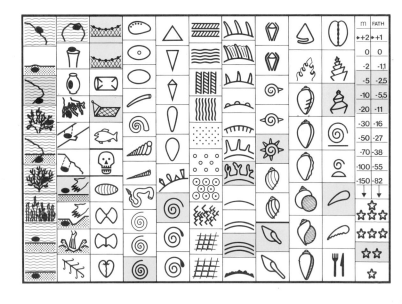

Murex pecten Lightfoot 1786   Muricidae
Venus comb murex

A real natural wonder, with its six lines of thin needle-like digitations.

Dist: Indian and Pacific provinces   100–150mm

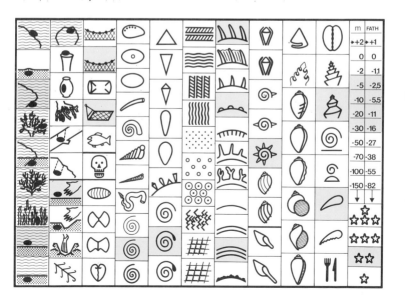

Phyllotonus regius Swainson 1821   Muricidae
Royal murex

Sought after by collectors because of its chocolate and strawberry colours.

Dist: Californian and Panamic provinces   60–120mm

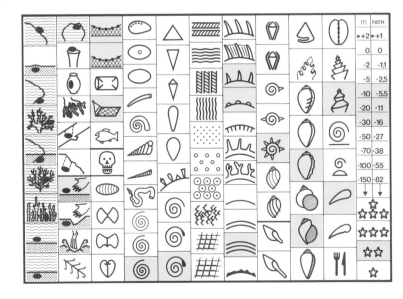

Columbarium pagoda Lesson 1831   Columbariidae
Common pagoda shell

This small family consists of fifteen species previously classified with the Muricidae. Some specimens have been found at a depth of a thousand metres.

Dist: Japanese province   50–80mm

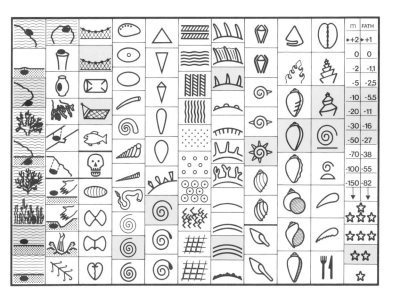

Rapa rapa Linnaeus 1758   Coralliophylidae
Papery rapa, Coral dwellers

As the name of the family implies, these molluscs live on or within coral
formations: they open a small hole through their host to keep in contact with
sea water.

Dist: Pacific province   40–80mm

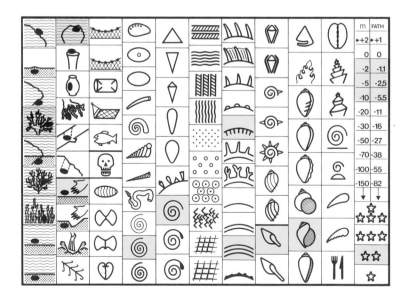

Latiaxis pilsbryi Hirase 1908   Coralliophylidae

This genus consists of some sixty species of extremely variable forms but all very attractive. Unlike the **Rapa**, they live in deep waters.

Dist: Japonic province   30–50mm

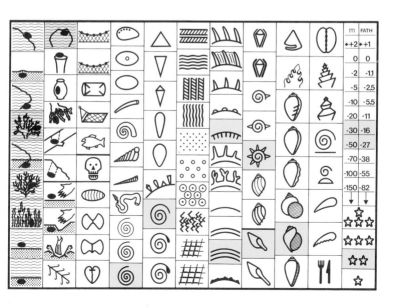

Various **Latiaxis** species   Coralliophylidae

These beautiful **Latiaxis** have been gathered in large quantities over the past few years by the Filipinos using their special nets devised entirely for the purpose of shell collecting. As a consequence, many species which were once rare are now within the reach of collectors.

Dist: Philippines, Japan   25–40mm

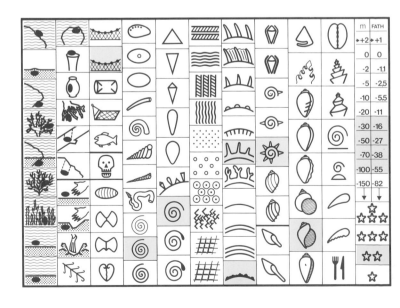

Thais planospira Lamarck 1822   Thaididae

This family has certain traits in common with the Muricidae: they live in colonies along rocky coastlines and feed on bivalves. They secrete a purple dye which the Indians of Central America used to dye their cotton.

Dist: Panamic province   35–60mm

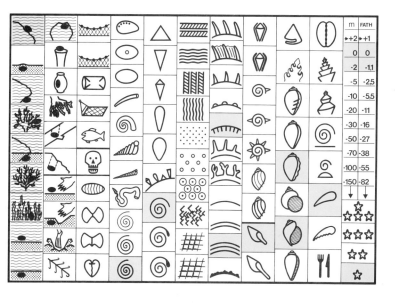

Drupa grossularia Röding 1798
Drupa morum Röding 1798
Drupa ricinus Linnaeus 1758

Flat, small shells living on corals in shallow waters.

Dist: Indian and Pacific provinces

Thaididae

20–30mm

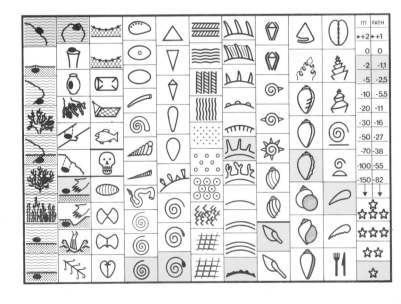

Babylonia areolata Link 1807   Buccinidae
Spotted babylon
Babylonia spirata Linnaeus 1758
Spiral babylon

The Buccinidae are widespread from the Arctic to the Tropics. They are predators or scavengers but do not bore holes.

Dist: Indian province   35–70mm

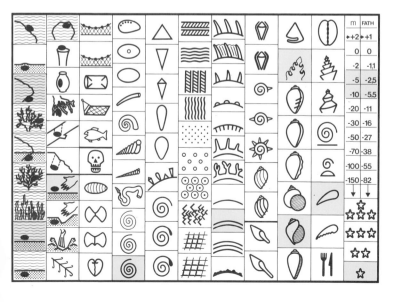

Opeatostoma pseudodon Burrow 1815   Buccinidae

This shell is immediately recognisable by its short digitation in the shape of a claw on the anterior. This genus is classified with the Fasciolaridae by some authorities.

Dist: Panamic province   25–70mm

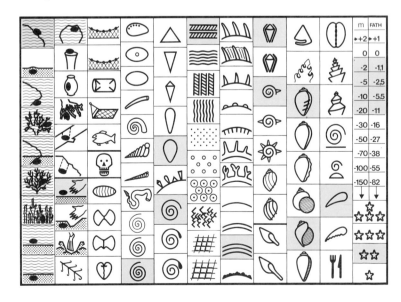

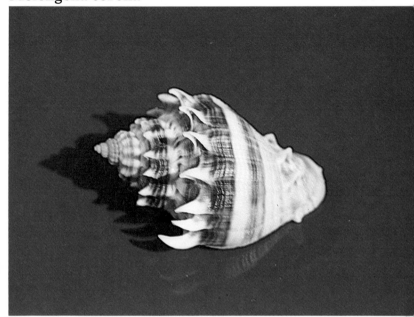

Melongena corona Gmelin 1791   Melongenidae
Florida crown conch

A carnivorous mollusc feeding on bivalves and gastropods.

Dist: Caribbean province   50–100mm

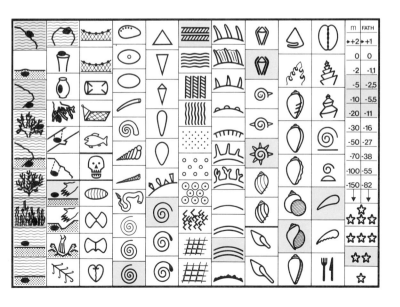

Syrinx aruanus Linnaeus 1758   Melongenidae
False trumpet

The largest of the gastropods; a specimen found at Malibu (California) in the spring of 1982 was 90cm long. Native tribes have used this shell for gathering water. The first eight spirals of the shell, secreted by the young animal, are all the same diameter.

Dist: South-west Australia   → 900mm

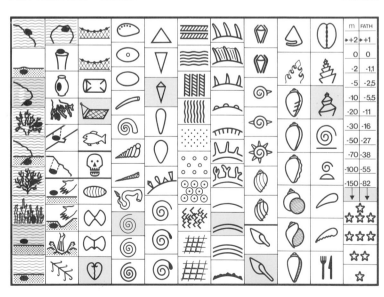

Busycon contrarium Conrad 1867   Melongenidae
Lightning whelk

The shells of this species are usually sinistral. A carnivorous mollusc which opens bivalves with the edge of its own shell.

Dist: Caribbean province   → 170mm

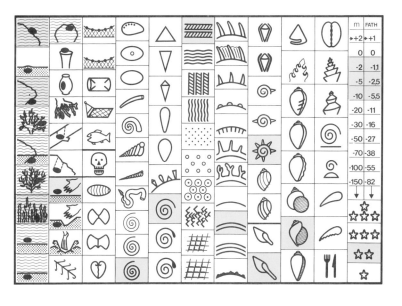

Fasciolaria tulipa Linnaeus 1758   Fasciolariidae
True tulip shell; Tulip spindle shell

Like all members of its family, **Fasciolaria tulipa** has a very thick, horny operculum. They are all carnivorous.

Dist: Caribbean province   →200mm

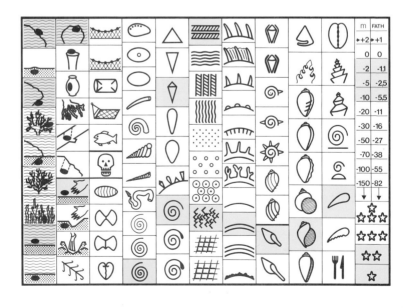

Fusinus colus Linnaeus 1758   Fasciolariidae
Distaff spindle

The genus **Fusinus** includes some fifty species which live on sandy seabeds; they move about in pairs.

Dist: throughout the Indian and Pacific provinces   → 200mm

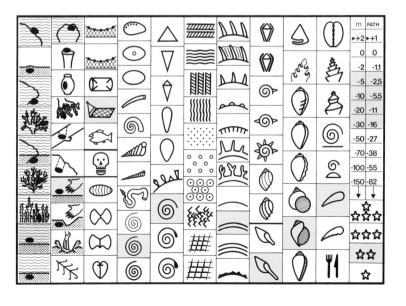

Oliva oliva Linnaeus 1758   Olividae
Olive

The genus **Oliva** has no operculum; the mantle completely covers the shell.
They can be gathered at low tide. The family consists of over 300 species, all
carnivorous: they live on fish, crabs and bivalves.

Dist: Indian and Pacific provinces   20–40mm

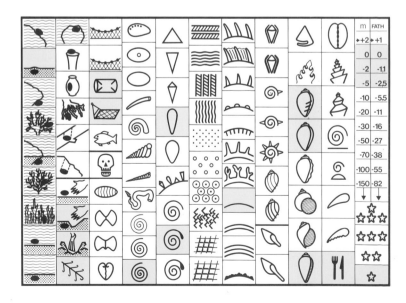

Oliva porphyria Linnaeus 1758   Olividae
Tent olive

This shell was collected by Europeans very early on, having been introduced from the New World by sailors and travellers. This is the largest of the olives.

Dist: Panama Gulf (Pedro Gonzales Island)   80–110mm

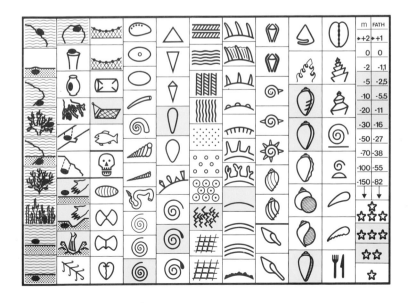

Mitra mitra Linnaeus 1758   Mitridae
Episcopal mitre

A carnivorous mollusc; some species have a poison-secreting gland with which
they kill their prey. There are about 600 species in the two provinces of the
Indo-Pacific. The majority live in the tidal areas, others have been dredged
from a depth of up to 1400m.

Dist: Indian and Pacific provinces   70–180mm

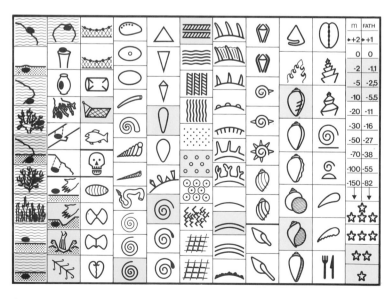

Mitra mitra Linnaeus 1758   Mitridae
Episcopal mitre

This section shows the columella (the real axis of the shell) and the columellar folds which run regularly along its length (see Figure 1, page 19).

Dist: Indian and Pacific provinces   70–180mm

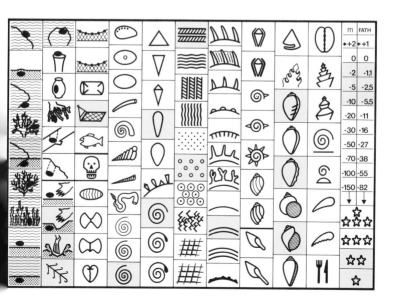

Vexillum regina Sowerby 1828   Mitridae
Vexillum taeniatum Lamarck 1811
Vexillum filiareginae Cate 1961 (form of **V. regina**)

Three very beautiful specimens found in the Philippines.

Dist: Indian and Pacific provinces   50–70mm

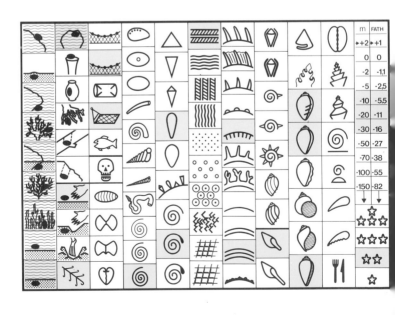

Vasum tubiferum Anton 1839   Vasidae
Imperial vase

About two dozen species are known, most of them quite common; they live in calm, shallow waters.

Dist: Philippines, New Guinea   70–120mm

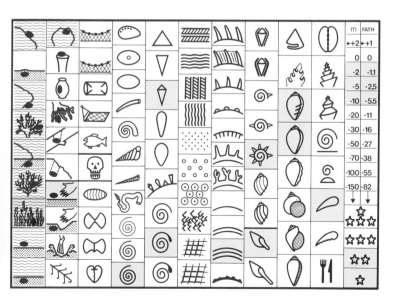

Turbinella pyrum Linnaeus 1758   Turbinellidae
Indian chank

The Turbinellidae can only be found in very few areas. Gathered in their
thousands in India and Sri Lanka, they are cut and mounted as jewellery. The
very rare sinistral specimens are worshipped in Hindu temples. **Cankh** is the
Hindustani word for shell.

Dist: Bengal Gulf and India   70–150mm

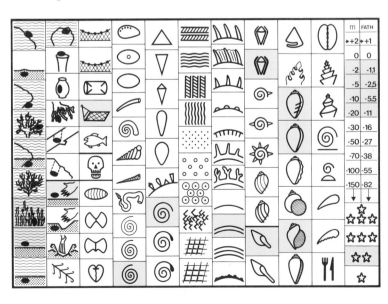

Harpa harpa Linnaeus 1758   Harpidae
Common harp

The Harpidae consist of only fourteen species, all carnivorous; they move about rapidly and are characterised by the autotomy phenomenon, or auto-amputation of the foot, which allows them to escape from predators.

Dist: throughout the Indian and Pacific provinces   50–80mm

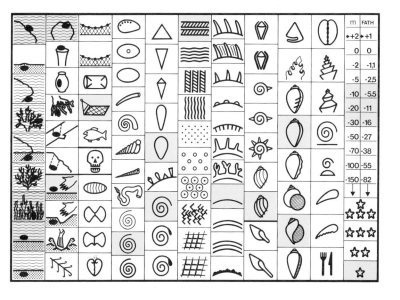

Harpa doris Röding 1798   Harpidae
Rosy harp

Relatively small, this uncommon harp is widely sought after because of its
beautiful red colour. The specimens found around the Cape Verde Islands are
more rounded and heavier than those from the African continent.

Dist: Angola, Ascension Island and Cape Verde Islands   30–80mm

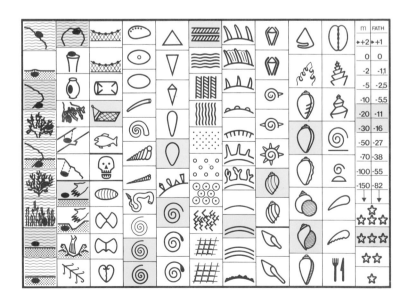

Amoria damonii Gray 1864   Volutidae
Damon's volute

A very beautiful shell with varying colours and patterns. There are four sub-species.

Dist: Australia   50–140mm

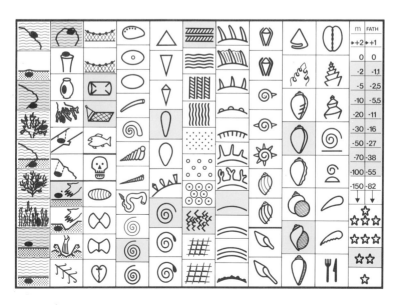

Lyria delessertiana Petit de la Saussaie 1842   Volutidae
Delessert's volute

This genus includes some twenty species which live on sandy, madreporic seabeds.

Dist: Madagascar, Comores, Seychelles   40–60mm

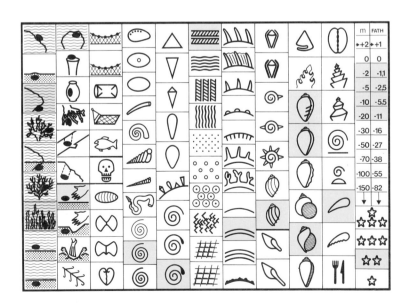

Voluta ebraea Linnaeus 1758   Volutidae
Hebrew volute

This shell owes its name to its characteristic patterns which resemble Hebraic writing. In 1964, Clench and Turner mentioned a gigantic specimen 260mm long.

Dist: Caribbean, Brazil   90–150mm

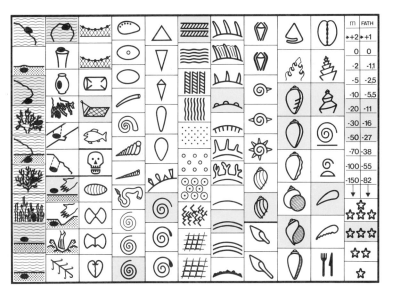

Cymbiola imperialis Lightfoot 1786   Volutidae
Imperial volute

A very beautiful shell, easily recognisable by its spines ranged on its spiral shoulder, parallel to the axis.

Dist: endemic to the Sulu Islands (Philippines)   125–200mm

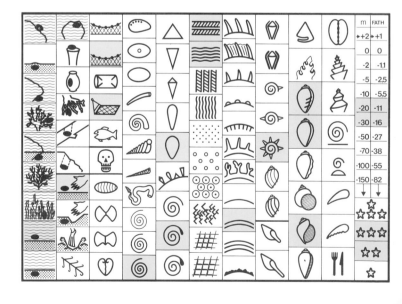

Melo melo Solander   Volutidae
Baler shell, Indian volute

All the shells of the genus **Melo** are quite small and smooth. The mollusc on the other hand is large and its body protrudes from the shell. The majority of these species come from West Africa (Senegal); the one illustrated was found at Pattaya, in the Gulf of Thailand.

Dist: Indian and Pacific provinces   → 250mm

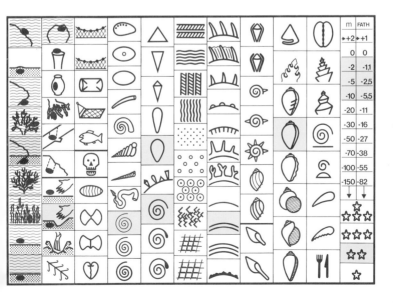

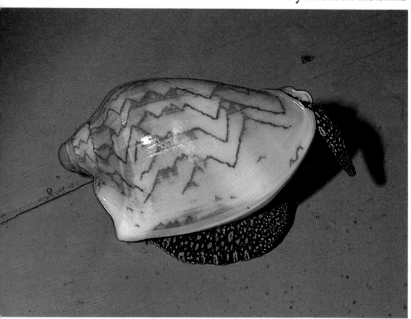

Cymbiola nobilis Lightfoot 1786   Volutidae

A beautiful volute, with variable patterns and colours. The specimen shown
here was found at a depth of 2m on a rock on a sandy and muddy seabed off the
coast of Pattaya (Gulf of Thailand) in August 1982; this increases considerably
the geographical distribution given by existing literature.

Dist: from Taiwan to Singapore and Thailand, to the south Philippines

80–200m

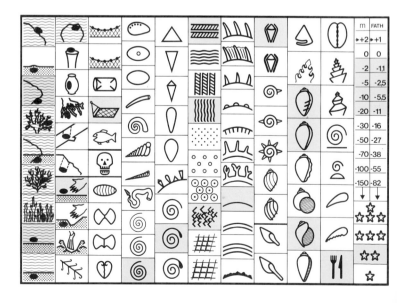

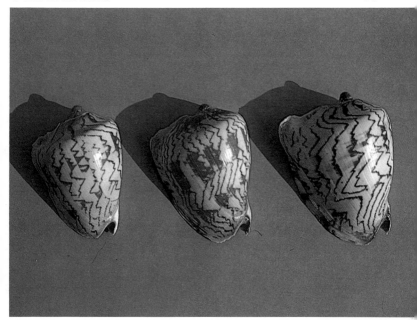

Cymbiola nobilis Lightfoot 1786   Volutidae
Noble volute

Although the pattern of this shell is constant, there are remarkable differences in size between adult specimens, according to the biotopes: 60–85mm in the shells found by the authors at Pattaya, over 180mm for those found in the Philippines. Very old specimens are characterised by a labial growth in the shape of a wing.

Dist: Gulf of Thailand to Singapore   60–180mm

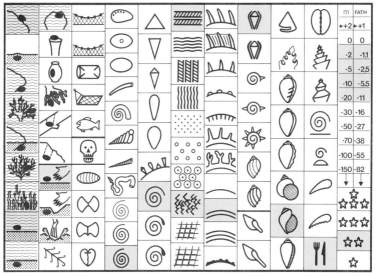

Cymbiola vespertilio Linnaeus 1758   Volutidae
Bat volute

There are over 200 known species of volutes. Most of them can be found in shallow waters; others, the rarest, have been dredged from depths of up to 1000m. The foot is very important (see the photograph) and often highly coloured. This specimen was photographed off the coast to the north of the Celebes Islands (Indonesia) in August 1981.

Dist: Indian and Pacific provinces   40–120mm

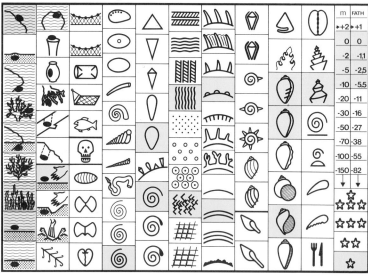

Cymbiola vespertilio Linnaeus 1758   Volutidae
Bat volute

Very variable in shape, size and colour, as shown by this plate. Sinistral specimens are not too expensive.

Dist: Indian and Pacific provinces   45–115mm

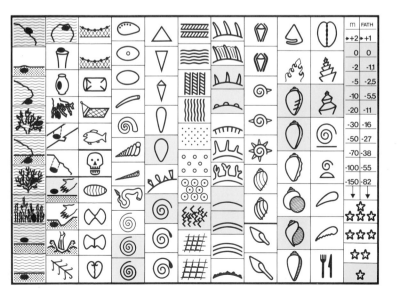

Marginella angustata Sowerby   Marginellidae

A family formed by about 600 species distributed in all the warm waters, but mainly along the coast of West Africa.

Dist: West African, South African, Indian and Pacific provinces   13–25mm

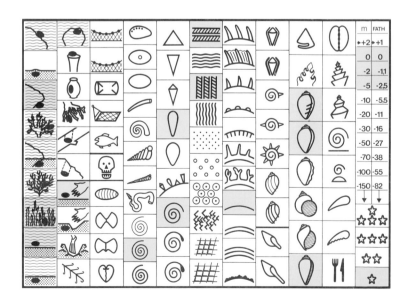

Persicula cingulata Dillwyn 1817   Marginellidae
Syn **M. lineata** Lamarck 1822
Persicula marginata Born 1778
Belted margin shell

Extremely variable forms, known under several synonyms. Very similar to **M. amygdala** Kiener and **M. cincta** Kiener.

Dist: West African province   15–25mm

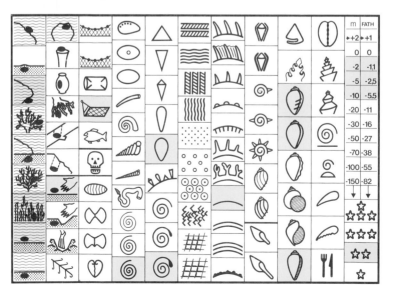

Marginella (**Afrivoluta**) **pringlei** Tomlin 1947

Marginellidae

The largest of the margin shells, it has long been considered to belong to the volutes. Rare until a few decades ago, it is now quite easy to obtain. It is characterised by a strong callosity on the shoulder near the aperture.

Dist: South African province   70–100mm

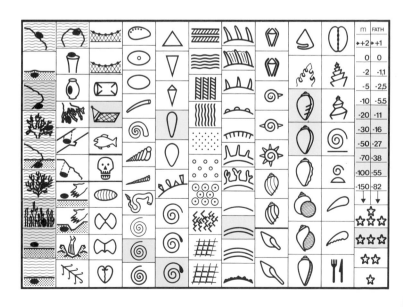

Turris babylonia (?) Linnaeus 1758   Turridae

A large family numbering several hundreds of species, many of which have an anal fissure like the Pleurotomariidae (not to be confused). Certain species also possess a poisonous gland similar to that of the cones.

Dist: throughout the Indian and Pacific provinces   50–70mm

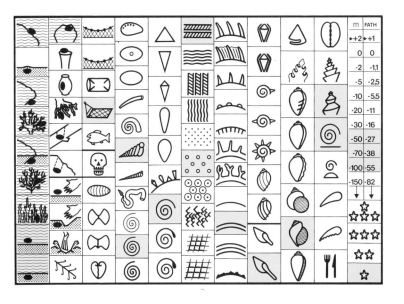

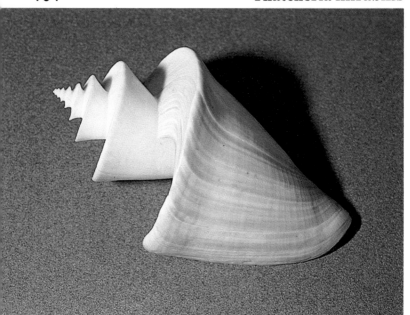

Thatcheria mirabilis Angas 1877　　　　　　　　Turridae
Japanese wonder shell

The largest of the Turridae. Its extraordinary shape makes it immediately recognisable from all other shells.

Dist: endemic to Japan　　　　　　　　　　　　　　75–100mm

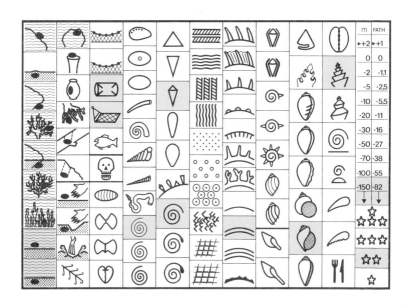

Conus ammiralis Linnaeus 1758   Conidae
Ammiralis cone

There are about 400 species of Conidae, the family with the most controversial nomenclature particularly at the level of sub-species. All cones are carnivorous and feed on worms and fish which they capture with specially adapted radular teeth.

Dist: Indian province   60–70mm

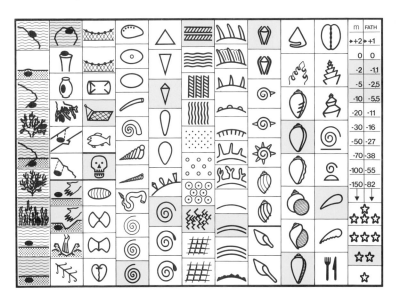

Conus barthelemyi Bernardi 1861   Conidae
Barthelemy's cone

A very beautiful cone; the young specimens are a deep red, becoming reddish-brown or black when adults.

Dist: Islands in the south of the Indian Ocean   40–90mm

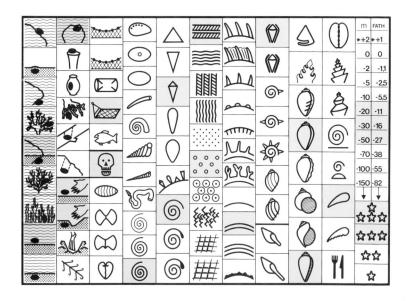

Cylinder bullatus Linnaeus 1758   Conidae
Bubble cone

A very beautiful cone both for its colour and its shining surface. A few years ago, local divers discovered a site (Raya Island near Phuket, south of Thailand) where **C. bullatus** abounded, especially the variety **C.b. articulata** Dautzenberg 1939, with its vivid orange aperture.

Dist: Indian and Pacific provinces   45–75mm

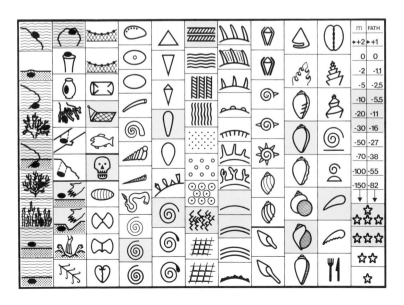

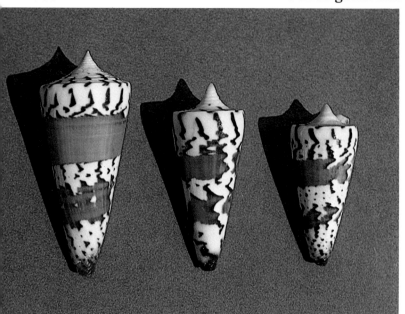

Leptoconus generalis Linnaeus 1767   Conidae

A species with very variable pattern and colour; it would be quite easy to gather a series of specimens with different origins and colours ranging from pale yellow to very dark brown.

Dist: throughout the Indian and Pacific provinces   45–75mm

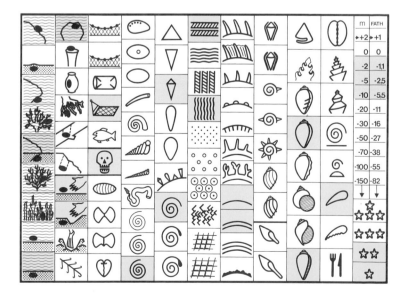

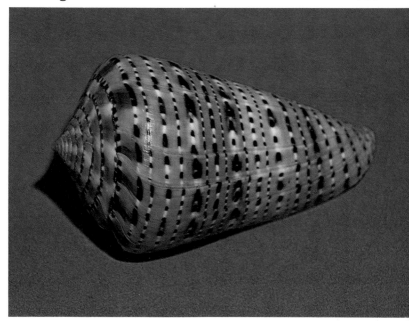

Conus genuanus Linnaeus 1758   Conidae

This attractive cone can easily be recognised by its transversal black bands broken by white spots. Relatively common, but perfect specimens are rarely found: the back is often corroded as it tends to protrude from the sand.

Dist: Senegal, Angola, Cape Verde Islands   30–70mm

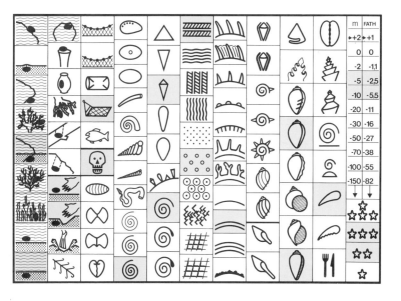

Gastridium geographus Linnaeus 1758   Conidae
Geography cone

One of the largest species of cones and one of the most dangerous: it has caused
many fatal accidents. The periostracum has parallel lines of bristles.

Dist: Indian and Pacific provinces   60–110mm

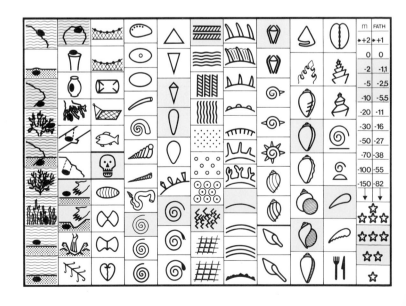

Leptoconus gloriamaris Chemnitz 1777   Conidae
Glory of the sea

For quite a long time only a few specimens of this splendid cone were known until a colony was discovered in the Solomon Islands. Each year, a few dozen specimens come from the Philippines, where they are fished from the depths with special nets.

Dist: Philippines and Pacific province   70–140mm

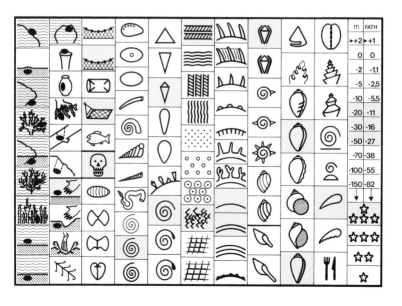

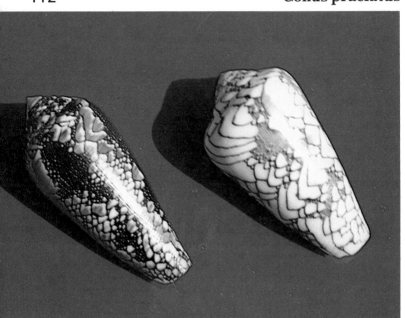

Conus praelatus Hwass in Bruguiere 1792   Conidae

A beautiful cone of variable colour, ranging from pale yellow to dark blue.

Dist: endemic to East Africa   50–65mm

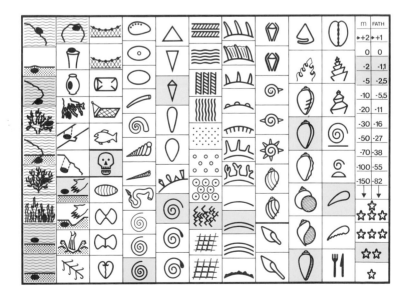

Conus pulcher Lightfoot 1786   Conidae

The largest of living cones. The best example is **c. prometheus** Hwass in Bruguiere 1792, and a sub-species was named **c. papilionaeus** by the same author.

Dist: West African province   300mm

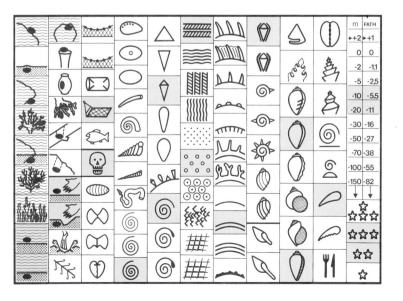

Conus sozoni Bartsch 1939   Conidae
Sozoni's cone

This elegant cone was fished out of the Florida waters by Greek sponge
fishermen. Other specimens are dredged by trawlers.

Dist: Caribbean province, only on the continental coasts   40–60mm

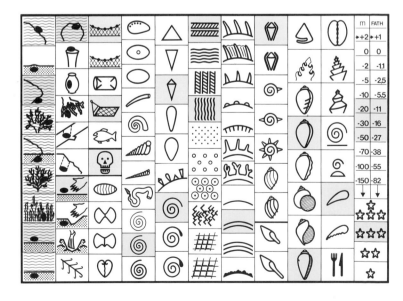

Darioconus textile Linnaeus 1758   Conidae
Textile cone

Known under various synonyms owing to its numerous variations: **verriculum**, **eumitus**, etc. Carnivorous, it feeds on other molluscs. Its poison is very strong.

Dist: throughout the Indian and Pacific provinces   70–110mm

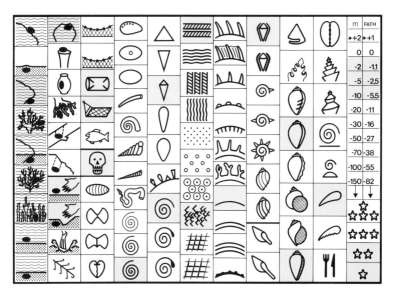

Terebra maculata Linnaeus 1758   Terebridae
Marlinspike

The largest of a family which numbers about 300 species. Its poisonous organ
resembles that of the cones. It leaves a large track in the sand.

Dist: throughout the Indian and Pacific provinces   150–250mm

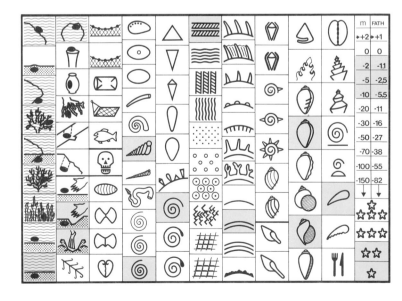

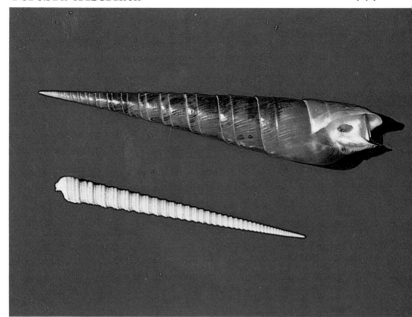

Terebra triseriata Gray 1834   Terebridae
Triseriate auger

Sought after by collectors as being the most elongated of the terebrids; it can
have over fifty spirals. Carnivorous, like all members of the family.

Dist: Indian and Pacific provinces   → 120mm

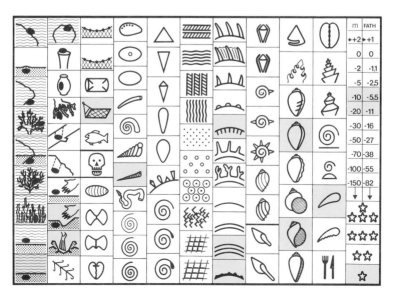

Hydatina albocincta van der Hoeven 1839   Hydatinidae

Shell extremely thin and fragile; the mantle completely envelops the shell. Lives in the sand. Carnivorous.

Dist: Indian, Pacific and Japonic provinces   25–40mm

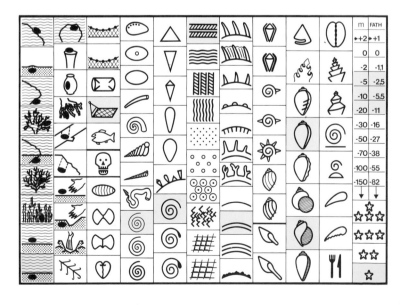

Hydatina physis Linnaeus 1758   Hydatinidae

The body of this mollusc is very well developed and can only partially retire inside the shell. It moves quickly by graceful undulations of the edge of its foot.

Dist: Indian and Pacific provinces   30–45mm

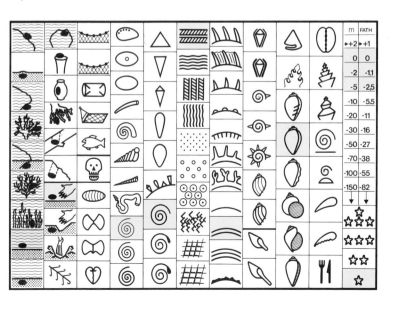

Pecten flabellum Gmelin 1791   Pectinidae

This family (scallops) numbers many hundreds of species, many of which are edible. Vividly coloured, the **P. flabellum** has no teeth on its hinge, only a thickening called **crura**.

Dist: West African province   40–60mm

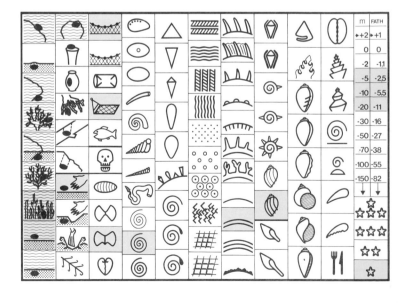

Pecten glaber Linnaeus 1758   Pectinidae

Its very variable colour enables one to collect a beautiful series of attractive tints. Like all Pectinidae, **P. glaber** rests or attaches itself by its right valve.

Dist: Mediterranean province   40–60mm

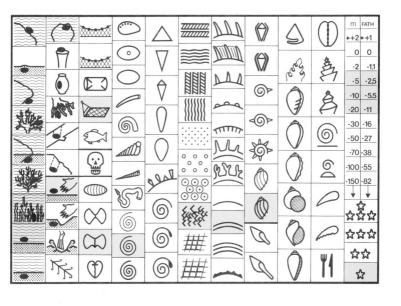

Chlamys speciosus Reeve 1853   Pectinidae

Endemic to Japan, this pecten is distinguished by the assortment of scales which cover the two valves. The right valve is monochrome (yellow or orange) while the left one displays brown patches on a plain ground.

Dist: Japonic province   40–60mm

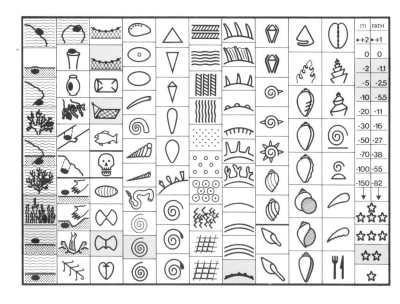

Spondylus americanus Hermann 1781   Spondylidae
Atlantic horny oyster

A beautiful bivalve, it anchors itself by its right valve to various supports, notably metal wrecks, usually in small groups of five or six individuals.

Dist: Caribbean province   70–150mm

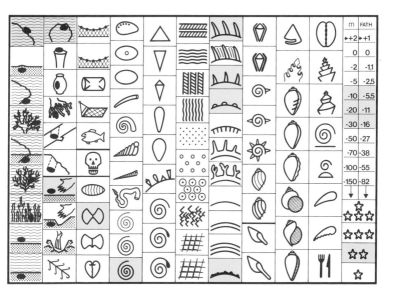

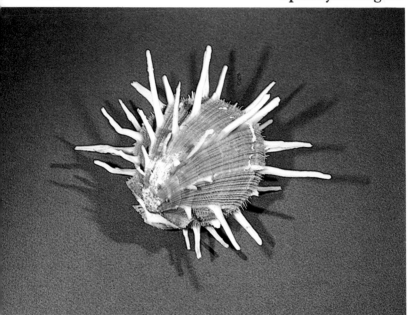

Spondylus regius Linnaeus 1758   Spondylidae
Regal Thorny oyster

Monochrome shell. Has a muscular ligament on the inside of each valve, like most bivalves, except that in this case it is greenish-brown in colour.

Dist: Indian and Pacific provinces   80–120mm

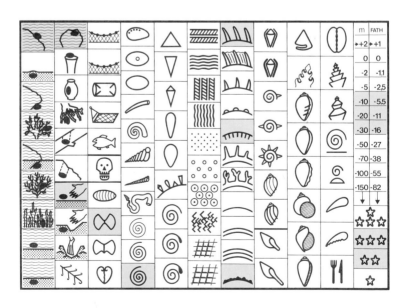

Lopha cristagalli Linnaeus 1758   Ostreidae
Coxcomb oyster

This oyster, with its saw-toothed aperture, has root-like growths which enable it to anchor itself to its supports, usually rocks.

Dist: Indian and Pacific provinces   50–90mm

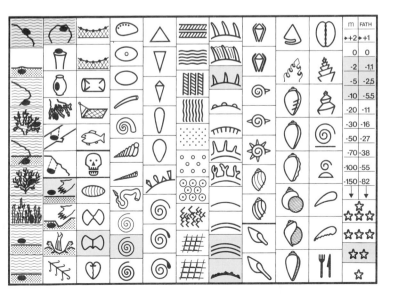

Chama lazarus Linnaeus 1758   Chamidae
Lazarus jewel box

This family numbers some twenty species. The chamas live anchored by their left valve on rocks and wrecks. As in the case of the spondyls, their shell is inequivalve. Their colours can be very variable.

Dist: Indian and Pacific provinces   50–100mm

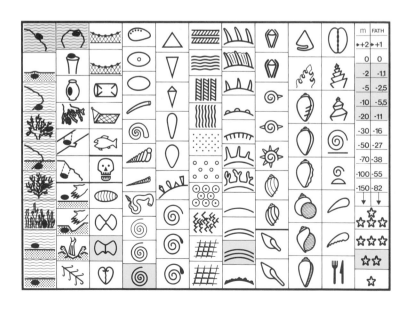

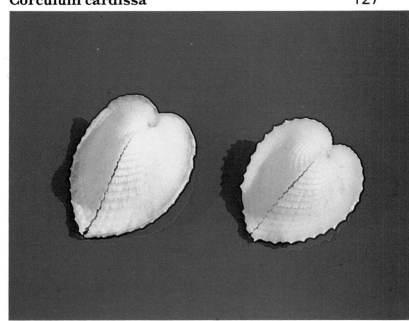

Corculum cardissa Linnaeus 1758

Cardiidae

Heart cockle

A very large family numbering seventeen genera living mainly in deep waters. They have a large foot and are very active. Edible.

Dist: Pacific province

40–60mm

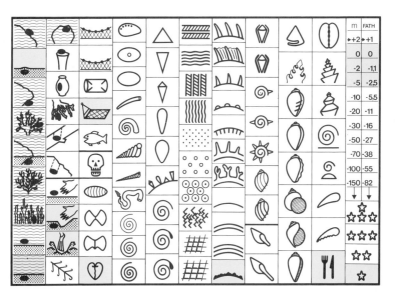

Cardium costatum Linnaeus 1758   Cardiidae
Ribbed cockle

A large bivalve with a thin shell embellished by hollow ribs.

Dist: West African province   80–100mm

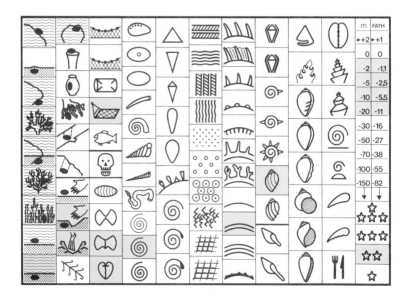

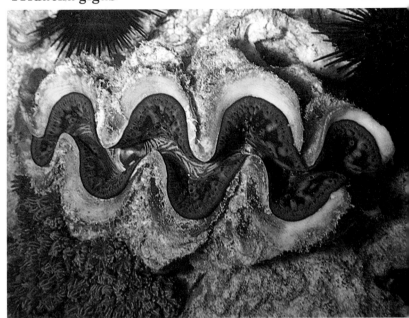

Tridacna gigas Linnaeus 1758
Giant clam

Tridacnidae

The largest bivalve in the world: it can reach up to 135cm long and each valve can weigh up to 230kg. This mollusc lives anchored to coral formations.

Dist: Indian and Pacific provinces

20–100cm

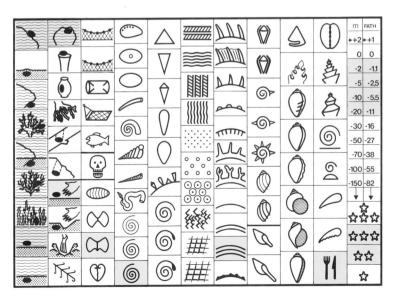

Hippopus hippopus Linnaeus 1758   Tridacnidae
Horseshoe clam, Bear's paw clam

Its vernacular name comes from the horseshoe-shaped surface (where the
byssal gape is) on the anterior of the shell. When young it is anchored to
madrepores, floating freely when adult.

Dist: Pacific province (western)   →40cm

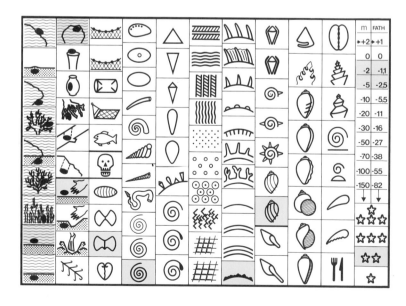

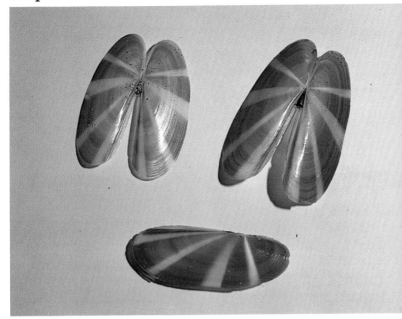

Siliqua radiata Linnaeus 1758   Cutellidae

The shells of this small family are characterised by the irregularity of the last few rings. The initial coils resemble those of the **Turritella**.

Dist: Indian and Pacific provinces   40–80mm

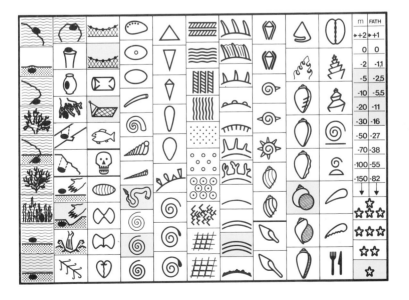

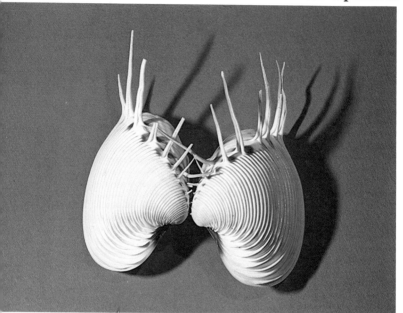

Pitar lupanaria Lesson 1830   Veneridae

There are over 400 species throughout the world, most of them edible. **P. lupinaria** is characterised by its long hollow spines ranged on the posterior margin of each valve.

Dist: Panamic and Peruvian provinces   50–80mm

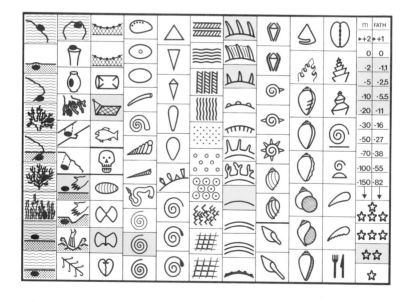

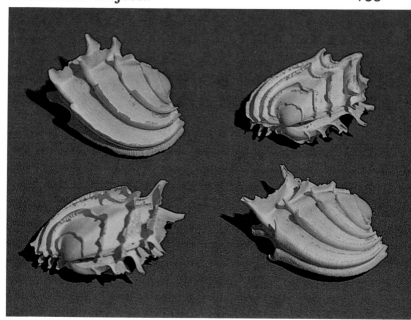

Callanaitis disjecta Perry 1811
Wedding cake venus

Veneridae

Relatively uncommon, and difficult to find in perfect condition. Fine parallel ribs on the anterior edge.

Dist: Southern Australia

50–70mm

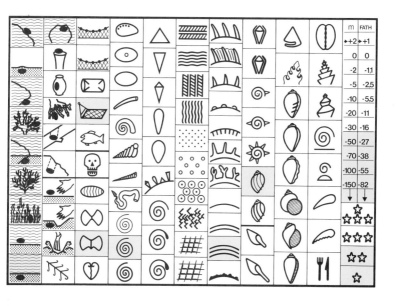

Venus lamellata Schumacher 1817   Veneridae

The interior of the shell is orangey-pink. The thin parallel ribs of the edge are
connected by small perpendicular ribs forming a reticulation.

Dist: Indian and Pacific provinces   40–50mm

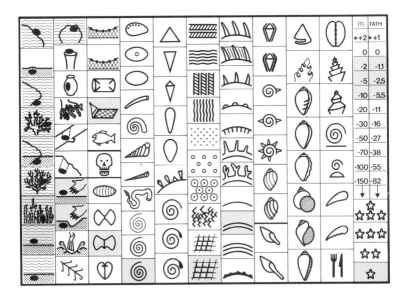

Glossus moltkianus Spengler 1783   Glossidae

This very beautiful shell is found at great depth in the Philippines. Off the coast of Japan, however, a sub-species can be found, **G.m. sanguinomaculata** Dunker, on sandy bottoms in shallow waters.

Dist: Philippines and Japonic provinces   40–60mm

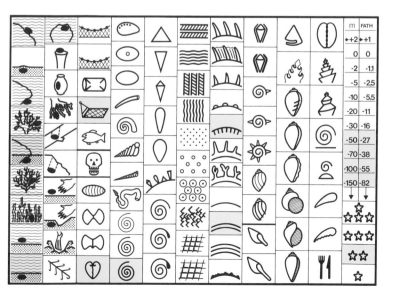

Barnea costata Linnaeus 1758   Pholadidae
Syn **Cyrtopleura costata** Linnaeus
Ribbed piddock, Angel wings

An elegant shell, fragile in appearance; it hollows out a hole equal to its length in wood, rocks or corals. It has an apophysis in the shape of a scoop.

Dist: Caribbean province, Florida   100–170mm

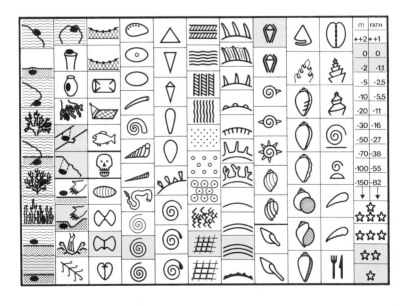

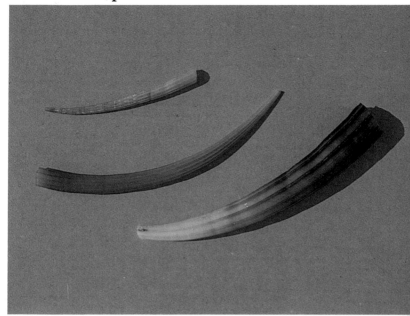

Dentalium elephantinum Linnaeus 1758   Dentalidae
Elephant's tusk shell

This family numbers some 300 species, living in the sand to a depth of 200m.
They have been used as currency, and magic powers were once attributed to
them (they were worn as amulets).

Dist: Indian and Pacific provinces   → 90mm

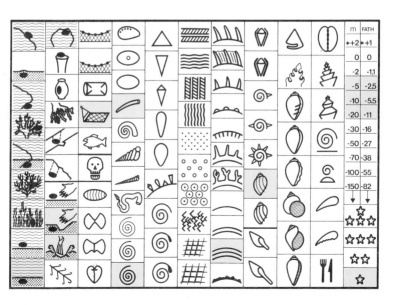

Argonauta argo Linnaeus 1758
Paper nautilus

Argonautidae

This is a pseudo-shell: it is secreted by the female as a nest for her eggs. The male is very small and has no shell. Argonauts are pelagic and live in all tropical seas. The shells float on the surface and can get stranded on a beach.

Dist: all warm seas

150–300mm

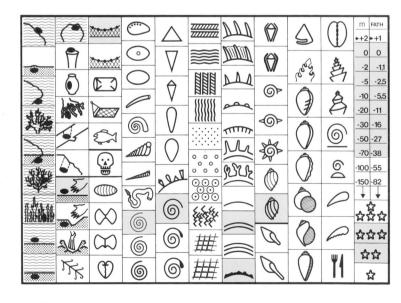

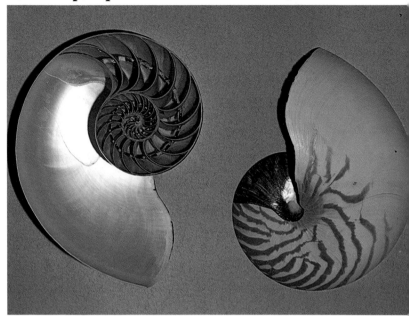

Nautilus pompilius Linnaeus 1758 — Nautilidae
Pearly nautilus, Chambered nautilus

The shell is divided into compartments which have hydrostatic purposes. The compartments are connected by a syphon and filled with a gas, variations in the volume of which affect the speed of ascent. Carnivorous.

Dist: Pacific province (western) — 150–250mm

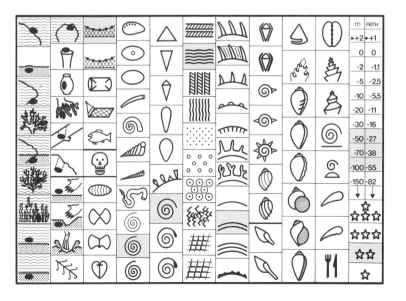

Spirula spirula Linnaeus 1758   Spirulidae

A shell presenting ventral rather than dorsal coiling as in the case of the **Nautilus pompilius**, with which it shares the same system of compartments and siphon. The animal is much larger than the shell and almost entirely covers it.

Dist: in all the seas   → 40mm

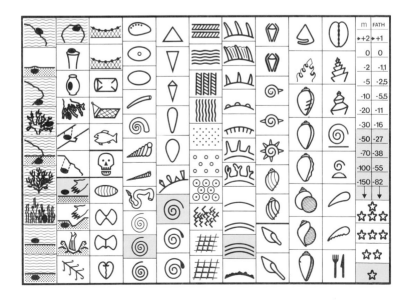

Katharina tunicata Wood 1815   Mopaliidae
Coat of mail

The specimen on the right has been decorticated to show the eight calcareous plaques characteristic of this class.

Dist: USA   40–70mm

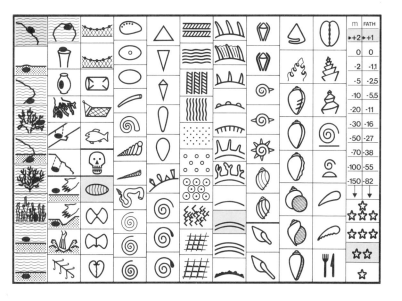

Papuina pulcherrima von Martens 1860   Carnaenidae
Green land snail

The English vernacular name derives from the normal colour of this shell;
yellow specimens are rare.

Dist: Melanesia   → 35mm

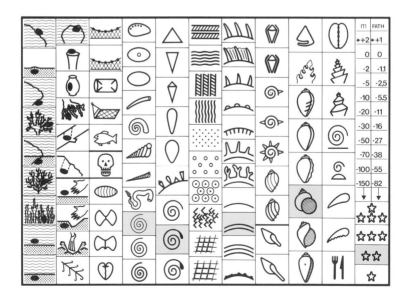

Xesta citrina Linnaeus 1758   Xanthonychidae
Syn **Naninia citrina** Linnaeus
Painted land snail

Shell with variable colouring. A very common tree species.

Dist: New Guinea, Celebes Islands   → 40mm

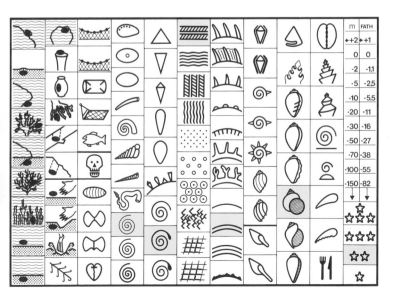

Helicostyla partuloides Broderip   Bradybaenidae

A beautiful shell with very variable pattern and colour.

Dist: Indian Ocean (east)   30–45mm

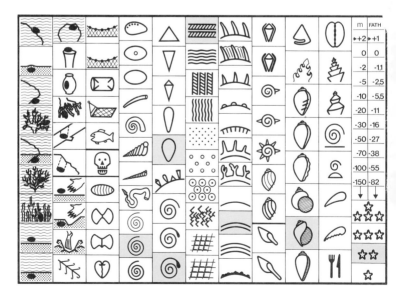

FURTHER INFORMATION

SOCIETIES AND CLUBS

In the Introduction, collectors' clubs and malacological societies are referred to; the advantages of belonging to them are:

—to meet other collectors
—to benefit from other collectors' experience and advice
—to have access to useful addresses
—to consult books and periodicals in their libraries
—to exchange shells
—to have your specimens identified
—to hear reports about a family or genus
—to take part in outings: to a beach to gather marine molluscs, or inland to gather terrestrial or freshwater molluscs
—to visit private collections and museums
—to see scientific films and transparencies
—to listen to travellers' tales and reports
—to be given advice on good collecting sites and all other details of climate, lodgings, budgeting, etc which will make your journey easier and more interesting, with malacology being one of the main interests.

Details of some collectors' clubs are as follows:

Club Français des Collectionneurs de Coquillages
Publications: *Xenophora*, 4 issues a year, about 64 pages.
50 Rue Richer, 75009 Paris, France.

Société Internationale de Conchyliologie
Publications: *Bulletin*, quarterly, about 70 pages.
Case postale 875, 1001 Lausanne, Switzerland.

Hawaiian Malacological Society*
Publications: *Hawaiian Shell News*, monthly, 160 pages.
PO Box 10391, Honolulu, 96816 Hawaii.

Société Belge de Malacologie
Publications: *Arion*, 6 issues a year, about 100 pages, local news.
Information, 4 issues a year, about 130 pages, international readership and a higher scientific level.
Exchange and information meetings: about 20, including 2 outings.
Avenue Mozart 52, 1190 Bruxelles, Belgium.

Club Conchylia
Am Steinern Kreuz 40, 6100 Darmstadt, West Germany.

Nederlandse Malacologische Vereniging
Secretary: H. P. M. G. Menkhorst, Weegbree 32, NL 2923 GM Krimpen a/d IJssel, Holland.

Belgische Vereniging Voor Conchyliologie
    Secretary: J. Wuyts, 82 Koningsarendl, B 2100 Deurne, Belgium.
British Shell Collectors' Club
    Secretary: David Feld, 14 Old Malling Way, Lewes, East Sussex,
    BN7 2EG.
Unione Malacologica Italiana
    Secretary and Administration: Via de Sanctis 73, 20145 Milano, Italy.

PUBLICATIONS

La Conchiglia (The Shell), edited by Ketty Nicolay.
    Not published by any specific club, this is a very handsome magazine
    with several colour illustrations. Bilingual Italian and English, 6 issues
    a year, 150 pages. Via C. Frederici 1, 00147 Roma, Italy.
Of Sea and Shore, edited by Thomas C. Rice. Quarterly in English. Varied
    contents; anecdotes, poems, etc. Illustrated with drawings and some
    colour plates. 4 issues of 64 pages. Port Gamble, Washington 98364,
    USA.

* In 40 years since its foundation, the membership of this club has risen to about
1600. It is the largest group of shell collectors in the world with members from over
70 countries.

BIBLIOGRAPHY

General books

French
Coquillages de Nouvelle-Calédonie et Mélanésie, S. Mayissian. Presse de
    Nouméa, 1974
Coquillages des Antilles, J. B. Lozet & C. Petron. Editions du Pacifique,
    1977
Coquillages du Monde Entier, O. & J. B. Lozet. Editions Maritimes et
    d'Outremer, 1978 (with price list)
Coquillages de Polynésie, B. Salvat & C. Rives. Editions du Pacifique, 1975
Coquillages des Côtes Atlantiques et de la Manche, Bouchet, Danrigal &
    Huyghens. Editions du Pacifique, 1978
Guide des Coquillages Marins, G. Lindner. Editions Delachaux et Niestlé,
    1976

English
The Encyclopedia of Shells, P. Dance. Blandford Press, 1974
Shells and Shell Collecting, P. Dance. Hamlyn Publishing Group, 1972
Rare Shells, P. Dance. University of California Press, 1969
Shell Collecting, P. Dance. Faber & Faber, 1966
The Shell Collector's Guide, P. Dance. David & Charles, 1976
Sea Shells of the World, A. G. Melvin. Charles E. Tuttle Company, 1966
    (with price list)
Marine Shells of the Pacific, W. O. Cernohorsky. Pacific Publications, 3 vols,
    1967–78
A Catalog of Dealers' Prices, edited by T. Rice. Of Sea and Shore
    Publication (published annually with price list)

Shells of the Western Pacific, T. Kira & T. Habe. Hoikusha, Japan, 2 vols 1962

Seashells of the World, J. M. Eisenberg. McGraw-Hill Book Co, 1981

Rare Shells of Taiwan, T. C. Lan. Published by the author, Taiwan, 1980

Shells of New Guinea and Central Indo-Pacific, A. Hinton. The Jacaranda Press, New York, 1972

Australian Shells, B. R. Wilson & K. Gillett. A Reed Book, Sydney, 1971

Seashells of Tropical West America, A. M. Keen. Stanford University Press, 1971

Kingdom of the Seashell, R. Tucker Abbott. Rutledge Books, New York, 1972

Sea Shells of the West Indies, M. Humfrey. Collins, London, 1975

Sea Shells of Southern Africa, B. Kensley. Maskew Miller Ltd, 1973

American Seashells, R. Tucker Abbott. Van Nostrand Co, 1974

Hawaiian Marine Shells, E. Alison Kay. Bishop Museum Press, Honolulu, 1979

Sea Shells of Sri Lanka, P. Kirtisinghe, Charles E. Tuttle Company, 1978

Van Nostrand's Standard Catalogue of Shells, edited by R. J. L. Wagner & R. Tucker Abbott. Van Nostrand Co, 1967

Specialist books

French

Cônes et Porcelaines de la Réunion, M. Veillard. Published by the author, 1976

Porcelaines Niger et Rostrées de Nouvelle Calédonie, J. M. Chatenay. Imprimeries Réunies de Nouméa, 1977

Porcelaines Mystérieuses de Nouvelle Calédonie, R. & G. Pierson. Published by the authors, 1975

Cônes de Nouvelle Calédonie et du Vanuatu, J. C. Estival. Editions du Pacifique, 1981

English

Living Cowries, C. M. Burgess. A. S. Barnes, New York, 1970

Cowries, J. Taylor & J. G. Walls. T. F. H. Publications, New York, 1975

Cone Shells of Thailand, A. J. da Motta & P. Lenavat. Published by the authors, Graphic Art, Bangkok, 1979

Cone Shells of the World, Marsh & Rippingale. The Jacaranda Press, New York, 1964

Conchs, Tibias and Harps, J. G. Walls. T. F. H. Publications, New York, 1980

Mitre Shells, P. Pechar, C. Prior & B. Parkinson. R. Brown & Associates, Australia, 1978(?)

Olive Shells of the World, R. F. Ziegler & H. C. Porreca. Published by the authors, New York, 1968

Contributions to the Study of Olividae, D. Greifeneder *et al.* Club Conchylia, Darmstadt, 1981

The Living Volutes, C. S. Weaver & J. E. du Pont. Delaware Museum of Natural History, New York, 1970

Cone Shells, J. G. Walls. T. F. H. Publications, New York, 1978(?)

The Murex Book, R. H. Fair. Published by the author, Honolulu, 1976

Murex Shells of the World, G. R. Radwin & A. D'attilio. Stanford University Press, California, 1976

A Review of the Volutidae, M. Smith. Borden Publishing Co, California, 1942

Many of the books in this bibliography are out of print: some may be found secondhand, the others in the libraries of museums or clubs.

GLOSSARY

aberrant: a specimen markedly different from other individuals of the same species

acid: (habitat) a habitat containing hydrogenated ions H^+

adductor: the muscle(s) in a bivalve which draws the two valves together and holds them closed

albinism: congenital anomaly consisting of the scarcity or absence of pigments

apex: the first coil of the spire, usually pointed

apical: pertaining to the apex

apophysis: natural projecting structure

axial: parallel with, or on, the longitudinal axis

axis: imaginary line through the shell apex to the anterior of the aperture; the columellar axis around which are coiled the whorls

benthic: pertaining to the seabed; benthic organisms are either free or anchored; they live in direct relationship with the seabed

biomass: quantity of living matter per surface or volume unit

biotope: habitat or geographical area in which the main climatic and biotic conditions are uniform

bivalve: lamellibranch molluscs whose shell consists of two pieces

byssus: bunch of fine, extremely resistant threads secreted by the foot and used by certain bivalves to anchor themselves to the substratum

callosity: thickening of the shell due to the accumulation of calcareous strata

canal, anal: _see_ **canal, posterior**

canal, anterior: notch in the aperture of gastropods; this canal, of varying length, is also called the siphonal canal as it allows the extrusion of the inhaling syphon of the mollusc

canal, posterior: notch through which excreta are expelled by gastropods (Figure 1, page 19)

canal, siphonal: _see_ **canal, anterior**

cancellate: ornament consisting of intersecting spiral and axial structures

cardinal: central tooth on hinge of bivalves

cephalopoda: class of molluscs having tentacles with suction pads growing from their heads

chitin: organic substance pertaining to the tegument of articulated invertebrates

chitinous: related to chitin

chlorophyll: green vegetal pigment which transforms carbon dioxide into organic matter by exploiting the energy of light

chromatophore: pigmented cell causing colour variations in certain animals, eg cephalopods

class: division of phylum

classification: taxonomy: kingdom – phylum – class – sub-class – order – family – genus – species

coelenterates: phylum of mainly marine animals with a body formed by two walls enveloping the digestive cavity and provided with urticating tentacles, eg corals, jellyfish

columella: central pillar of gastropod shells formed by the inner part of the whorls and around which the whorls coil

columella fold: ridges winding round the columella (eg the mitras)

columellar: related to the columella

conchiolin: an organic substance similar to chitin, also part of the shell

conchology: the science concerned with studying shells

detrivorous: feeding on animal wastes and detritus

dextral: clockwise coiling of the whorls towards the apex (with apex upwards, aperture facing the observer)

digitation: finger-like projections

dimorphism (sexual): the sum total of the characteristics which distinguish the two sexes

dorsal: in the direction of hinge region of bivalves, or uppermost side of certain gastropods, opposite the aperture

ecology: the study of the biotope in which organisms live and reproduce themselves as well as the relationship between such organisms and their surroundings

effluent: waste waters generally disposed of in town sewers

endemic: a species geographically limited to one area or one island

equivalve: a lamellibranch with identical valves; the opposite is inequivalve, eg a pecten

euphotic: constituting the upper layers of a mass of water subject to the action of light, from the surface to a depth of 100m; marine vegetal life takes place within these layers

family: subdivision of an order; it includes living organisms generally close in form or genealogy

fasciole: spiral band or groove produced by the growth of the shell

filtering: feeding on the plankton filtered from sea water

funiculus: a ripple in a gastropod's callus penetrating into the umbilicum

fusiform: spindle-shaped, tapering towards the apex and the base (gastropods)

gastropods: a class of molluscs which move about on a large, muscular foot; includes marine, freshwater and terrestrial species

genus: subdivision of a family; each genus numbers several species

gills: respiratory organs characterising crustaceans, molluscs and fishes

gorgonias: warm-water animals forming arborescent colonies (order of Octocoralliarians)

haemocyanine: a respiratory pigment containing copper which is present in the blood of molluscs and crustaceans

height: in gastropods, the distance between the apex and the siphonal canal; in bivalves the distance between the apex and ventral margin

herbivorous: feeding on vegetable matter

hinge: series of ligaments and interlocking teeth which allow the articulation of the valves of a lamellibranch

holotype: an individual designated by an authority as representative of a newly described species

imperforate: lacking an umbilicus

inequivalve: a lamellibranch with dissimilar valves

intertidal: the area between high- and low-water marks

irisation: to become rainbow-coloured, irridescent, as mother-of-pearl

labial: referring to the lip

lamellibranch: class of molluscs with calcareous shell, two valves, reduced head and lamellar gills (bivalves)

larva: juvenile form with independent lifestyle

last (or body) whorl: the last-formed whorl of a gastropod shell within which most of the animal is contained

lateral teeth: teeth on hinge plate situated on either side of cardinal tooth of a bivalve, or on either side of the central teeth on the radula of a gastropod

ligament: elastic, corneous structure placed below the hinge and joining the two valves of a lamellibranch; it causes the valves to open when the adductor muscle, which holds them closed, is relaxed

littoral: from high-tide mark to shallow waters

maculated: irregularly spotted

madrepores: order of celenterates including isolated or colonial polyps

mangrove: intertropical vegetation characterised by trees which sink their roots into coastal swamps, even when brackish

mantle: part of a mollusc which secretes the shell and covers the animal's organs

mantle cavity: cavity containing gills and visceral mass

melanism: congenital anomaly resulting from the excessive accumulation of melanin pigments making an animal look very dark, or even black

necton: group of marine animals endowed with locomotive organs (as opposed to plankton)

nodule: knot-like swelling on a shell

operculum: corneous or calcareous plaque which molluscs use, among other things, to close their shells

outer lip: outer margin of aperture from suture to base of columella

ovigerous: bearing eggs; each ovigerous capsule of the Cypraeidae contains several dozen embryos

pallial: forming part of the mantle

pallial line: impressed line on each valve of a bivalve corresponding with line of attachment of the ventral margin of the mantle muscle

pallial sinus: embayment of pallial line indicating former attachment of siphonal retractor muscle

pelagic: animals which live suspended in the water far from the seabed

periostracum: proteic coating or skin which covers the shell of many gastropods

peristome: margin of the aperture of a shell

phylum: one of the basic divisions of the vegetal and animal worlds

plankton: marine organisms transported by sea currents; phytoplankton are vegetal organisms, zooplankton are animal organisms

predator: organism feeding on other living organisms having previously killed them

proboscis: trumpet-shaped organ of certain gastropods, situated between the two tentacles

radial: shell ornamentation directed from umbo towards margins of bivalves

radula: a series of several teeth, often hundreds, growing on an often very long ribbon-like organ (gastropods); the examination of the radula often leads to the identification of difficult specimens

scaphopoda: marine molluscs characterised by a tubular shell in the shape of an elephant's tusk

septum: transverse calcareous plate dividing the interior of certain gastropod and bivalve shells

sessile: describes organisms anchored to the substratum, for instance bivalves attached by a byssum, and certain gastropods

sinistral: describes gastropods with the opening on the left of the columella, the whorls being coiled anti-clockwise towards the apex (with apex upwards, aperture facing the observer)

siphon: tubelike organ of gastropod and similarly shaped extension of the mantle of a bivalve; can be inhalant, ie channelling water to the animal, or exhalant, rejecting internal water

species: subdivision of a genus; it includes those individuals born of the same parents or having similar characteristics which are fertile in respect of members of the same species, but sterile in respect of those of different ones

spire: coiling of the shell before the body whorl

stria: incised groove or furrow on the shell; the growth striae, or lines, indicate the regular secretion of calcium carbonate layers

substratum: support or area of attachment

suture: spiral line on shell where whorls join

symbiosis: a mutually beneficial association between two organisms

taxonomy: the study of the characteristics upon which the classification of living organisms is based

tidal area: that part of the littoral alternately covered and uncovered by the tides; also called intertidal zone

tubercle: small, variously shaped elevation on shell surface

type series: all of the individuals of a new species gathered in the same area and forming the basis for the authority's description of the series

umbilicus: the hollow columellar axis in certain gastropods where the inner surfaces of the coils are not welded

umbo (or beak): a small protuberance marking the first part of the shell to be formed

univalve: one-piece shell

varix: a usually axial growth, often very prominent and rounded (a former outer lip margin not covered by successive growth)

veliger: planktonic mollusc larva

ventral: the margin of a bivalve opposite to the ligament; also the apertural side of a gastropod shell

visceral mass: the whole of the visceral organs

INDEX OF LATIN NAMES
(by family)

Siliquariidae:
Siliquaria anguina 21

Spirulidae:
Spirula spirula 140

Spondylidae:
Spondylus americanus 123
Spondylus regius 124

Strombidae:
Lambis crocata 32
Lambis scorpius 31
Strombus aurisdianae 24
Strombus canarium 25
Strombus gallus 26
Strombus gigas 27
Strombus lentiginosus 28
Strombus listeri 29
Strombus sinuatus 30
Tibia fusus 33

Terebridae:
Terebra maculata 116
Terebra triseriata 117

Thaididae:
Drupa grossularia 74
Drupa morum 74
Drupa ricinus 74
Thais planospira 73

Tonnidae:
Malea ringens 62
Tonna perdix 61

Tridacnidae:
Hippopus hippopus 130
Tridacna gigas 129

Trochidae:
Angaria melanacantha 11
Bathybembix argenteonitens 10
Cittarium pica 7
Clanculus pharaonius 8
Maurea tigris 9
Trochus niloticus 6

Turbinellidae:
Turbinella pyrum 88

Turbinidae:
Astraea heliotropium 16
Guildfordia yoka 17
Phasianella species 15
Turbo marmoratus 12
Turbo petholatus 14
Turbos divers (opercula) 13

Turridae:
Thatcheria mirabilis 104
Turris babylonia 103

Turritellidae:
Turritella terebra 19

Vasidae:
Vasum tubiferum 87

Veneridae:
Callanaitis disjecta 133
Pitar lupinaria 132
Venus lamellata 134

Volutidae:
Amoria damonii 91
Cymbiola nobilis 96
Cymbiola vespertilio 98 99
Lyria delessertiana 92
Melo melo 95
Voluta ebraea 93
Voluta imperialis 94
Voluta nobilis 97

Xanthonychidae:
Xesta citrina 143

Xenophoridae:
Stellaria solaris 35
Xenophora pallidula 34

INDEX

Numbers given refer to colour plates

WILD HERBS: A FIELD GUIDE
J. de Sloover & M. Goossens

Whether used as a practical identification guide in the field, or for armchair browsing, this book offers a great deal of information succinctly presented. The herbs in the 144 stunning colour plates are grouped by colour. To identify a plant, simply open the guide at the pages bordered by the colour corresponding to that of the flower and you will soon find the plant itself. This ingenious system is time-saving, and the close proximity of illustrations of similarly coloured flowers helps to avoid misidentification.

For the purposes of this book, a herb is defined as a useful plant, one which is used to cure, to feed, to flavour dishes, to dye wool, or for any other specific aim. Pictograms presented alongside each colour plate summarise other properties – aromatic, medicinal and culinary, which parts are efficacious, when the herb is at its prime, where it grows, when it flowers. This at-a-glance information is supplemented by useful appendices, a glossary and notes on further reading.

MUSHROOMS & TOADSTOOLS: A COLOUR FIELD GUIDE
U. Nonis

Anyone interested in collecting mushrooms, whether to study them scientifically or simply to enrich their everyday diet with their nutritional value, will find this book an invaluable guide. It is based on a new descriptive system: each of the 168 colour photographs shows specimens in their natural habitat and is accompanied by a pictogram giving identification – at a glance – of their principal characteristics, combining a wealth of information with simplicity of presentation. Identification is further aided by the colours on the margins of the pages which reflect those of the fungi.

An introduction describes the main genera, their habitat, dangerous or valuable properties, and directions for collecting and growing them. Further Reading, Etymology of Scientific Terms and indexes contribute to the unique value of this guide.

MINERALS & GEMSTONES: AN IDENTIFICATION GUIDE
G. Brocardo

A completely new way to approach the mineral world is here offered to the experienced collector and the beginner. The 156 splendid colour photographs are accompanied by a brief description and a pictographic table which provides, through easily recognisable symbols, the available information necessary for identification and classification. The key to the symbols is printed on a bookmark. The margins of the pages are coloured to reflect those of the minerals themselves and aid quick identification.

The colour plates are preceded by extensive information on how to recognise and collect minerals, their origin and formation, their structure and properties, the classification, and how to prepare them for preservation. A glossary, bibliography and indexes complete the volume and add to its value as an indispensable guide for all collectors.

MOUNTAIN FLOWERS: A COLOUR FIELD GUIDE
S. Stefenelli

Thousands of enchanting flowers grow on the mountain slopes of Europe, and this book will prove an informative and useful guide for those wishing to discover more about them, appreciating their beauty and understanding the need for their conservation.

Recognition is easy with the aid of the 168 splendid colour photographs. To identify a flower, simply match the colour of the flower to the corresponding colour section and your task becomes easy. Once it has been clearly identified, the pictograms which accompany the plates will enable the beginner and the serious botanist to discover at a glance all the other interesting facts about the flower. A bookmark showing the key to the pictogram, a section on habitat, a glossary of pharmaceutical terminology, a bibliography and two indexes add to the value of the book.

FRESHWATER AQUARIUM FISH: A COLOUR GUIDE
J. P. Gosse

Fish, the most ancient vertebrates in the world, present an amazing diversity of form, colour and mode of life. New techniques for underwater exploration and means of transport and storage have fostered an ever-increasing knowledge of their ways and, with it, growing popularity for aquariums. Here you will find the answers to most of the questions you are likely to ask. Can the fish of your choice be placed in a tank where other species already thrive? What sort of food will it need? How does it reproduce? What temperature should the water be kept at? The answers are condensed in the pictograms which accompany each plate; these schematic little drawings, fully explained in the text, allow you to see at a glance what temperament and biological habit characterise your fish. The clarity of the 144 colour photographs is a precise guide to identification.

BUTTERFLIES: A COLOUR FIELD GUIDE
M. Devarenne

Many of the butterflies found in our gardens and our countryside are threatened by a hostile environment and the effects of pesticides and insecticides. The aim of this guide is to offer the reader a wealth of information about the different species, and by doing so encourage his interest in helping to conserve these 'winged jewels'.

Each species is photographed in colour in its natural environment, while a pictogram aids identification by detailing, in visual form, its major characteristics. An introduction gives a detailed explanation of the symbols used, information about the butterfly's life cycle and details of the characteristics of the different familes. The text is complemented by a glossary, bibliography and index.

BIRDS OF EUROPE: A COLOUR FIELD GUIDE
L. Gonnissen & G. Mornie

Both beginners and experts will find this book an invaluable guide to bird identification. Learning the names of birds and discovering a variety of details regarding their habits is a fascinating hobby with several surprises in store. The task of identification and classification is made easier by the colour photographs of each species and their accompanying pictograms. These schematic drawings show at a glance the geographical distribution and biotope of the bird, together with its social behaviour, nesting and reproductive habits and the protection the species is offered by man. So often it is a bird's behaviour which identifies it from other species similar in appearance and the pictograms give instant clarification. The 144 species represented are all to be found in western Europe, some within the boundaries of gardens and public parks.

HOUSEPLANTS: A COLOUR GUIDE
L. Cretti & G. Barnabé Bosisio

Plants give interest and a lived-in atmosphere to every room, and brighten up porches, balconies and conservatories. Whilst certain houseplants are extremely hardy and thrive on minimal care and attention, others are more delicate and require a carefully controlled environment. The 144 colour illustrations included in the guide are accompanied by a brief description of each plant, providing precise details for identification and classification. A pictogram indicates the environment and type of care each one requires. A quick glance can provide information on the size of pot needed, light and humidity requirements, intensity and frequency of watering, methods of propagation and other useful tips. A bibliography and index complete this book which forms an indispensable guide for houseplant owners.